James Whitcomb Riley, Bill Nye

Nye's and Riley's Railway Guide

James Whitcomb Riley, Bill Nye

Nye's and Riley's Railway Guide

ISBN/EAN: 9783744679138

Printed in Europe, USA, Canada, Australia, Japan

Cover: Foto ©berggeist007 / pixelio.de

More available books at **www.hansebooks.com**

NYE AND RILEY'S

RAILWAY GUIDE

BY

EDGAR W. NYE

AND

JAMES WHITCOMB RILEY

Illustrated by Baron DeGrimm E. Zimmerman
Walt. McDougall and others

CHICAGO
THE DEARBORN PUBLISHING COMPANY
88 AND 90 LA SALLE STREET
NEW YORK   1888   SAN FRANCISCO

Why it was done.

What this country needs, aside from a new Indian policy and a style of poison for children which will be liable to kill rats if they eat it by accident, is a Railway Guide which will be just as good two years ago as it was next spring — a Railway Guide if you please, which shall not be cursed by a plethora of facts, or poisoned with information — a Railway Guide that shall be rich with doubts and lighted up with miserable apprehensions. In other Railway Guides, pleasing fancy, poesy and literary beauty, have been throttled at the very threshold of success, by a wild incontinence of facts, figures, asterisks and references to meal stations. For this reason a guide has been built at our own shops and on a new plan. It is the literary *piece de resistance* of the age in which we live. It will not permit information to creep in and mar the reader's enjoyment of the scenery. It contains no railroad map which is grossly inaccurate. It has no time-table in it which has outlived its uselessness. It does not prohibit passengers from riding on the platform while the cars are in motion. It permits everyone to do just as he pleases and rather encourages him in taking that course.

The authors of this book have suffered intensely from the inordinate use of other guides, having been compelled several times to rise at 3 o'clock A.M. in order to catch a car which did not go and which would not have stopped at the station if it had gone.

They have decided, therefore, to issue a guide which will be good for one to read after one has missed one's train by reason of one's faith in other guides which we may have in one's luggage.

Let it be understood, then, that we are wholly irresponsible, and we are glad of it. We do not care who knows it. We will not even hold ourselves responsible for the pictures in this book, or the hard-boiled eggs sold at points marked as meal stations in time tables. We have gone into this thing wholly unpledged, and the man who gets up before he is awake, in order to catch any East bound, or West bound, North bound, South bound, or hide-bound train, named in this book, does himself a great wrong without in any way advancing our own interests.

The authors of this book have made railroad travel a close study. They have discovered that there has been no provision made for the man who erroneously gets into a car which is side-tracked and swept out and scrubbed by people who take in cars to scrub and laundry. He is one of the men we are striving at this moment to reach with our little volume. We have each of us been that man. We are yet.

He ought to have something to read that will distract his attention. This book is designed for him. Also for people who would like to travel but cannot get away from home. Of course, people who do travel, will find nothing objectionable in the book, but our plan is to issue a book worth about

$9 charging only fifty cents for it and then see to it that no time tables or maps which will never return after they have been pulled out once, shall creep in among its pages.

It is the design of the authors to issue this guide annually unless prohibited by law and to be the pioneers establishing a book which shall be designed solely for the use of any body who desires to subscribe for it.

BILL NYE.

JAMES WHITCOMB RILEY.

P. S.—The authors desire to express their thanks to Mr. Riley for the poetry and to Mr. Nye for the prose which have been used in this book.

Contents

Where he First met his Parents.

Last week I visited my birthplace in the State of Maine. I waited thirty years for the public to visit it, and as there didn't seem to be much of a rush this spring, I thought I would go and visit it myself. I was telling a friend the other day that the public did not seem to manifest the interest in my birthplace that I thought it ought to, and he said I ought not to mind that. "Just wait," said he, "till the people of the United States have an opportunity to visit your tomb, and you will be surprised to see how they will run excursion trains up there to Moosehead lake, or wherever you plant yourself. It will be a perfect picnic. Your hold on the American people, William, is wonderful, but your death would seem to assure it, and kind of crystallize the affection now existing, but still in a nebulous and gummy state."

A man ought not to criticise his birthplace, I presume, and yet, if I were to do it all over again, I do not know

whether I would select that particular spot or not. Sometimes I think I would not. And yet, what memories cluster
about that old house! There was the place where I first
met my parents. It was at that time that an acquaintance
sprang up which has ripened in later years into mutual
respect and esteem. It was there that what might be
termed a casual meeting took place, that has, under the
alchemy of resistless years, turned to golden links, forming a pleasant but powerful bond of union between my
parents and myself. For that reason, I hope that I may
be spared to my parents for many years to come.

Many memories now cluster about that old home, as I
have said. There is, also, other bric-a-brac which has accu-
mulated since I was born there. I took a small stone from
the front yard as a kind of memento of the occasion and the
place. I do not think it has been detected yet. There
was another stone in the yard, so it may be weeks before
any one finds out that I took one of them.

How humble the home, and yet what a lesson it should
teach the boys of America! Here, amid the barren and
inhospitable waste of rocks and cold, the last place in the
world that a great man would naturally select to be born in,
began the life of one who, by his own unaided effort, in
after years rose to the proud height of postmaster at Laramie City, Wy. T., and with an estimate of the future that
seemed almost prophetic, resigned before he could be characterized as an offensive partisan.

Here on the banks of the raging Piscataquis, where
winter lingers in the lap of spring till it occasions a good
deal of talk, there began a career which has been the wonder
and admiration of every vigilance committee west of the
turbulent Missouri.

There on that spot, with no inheritance but a predispo-
sition to baldness and a bitter hatred of rum; with no
personal property but a misfit suspender and a stone-bruise,

began a life history which has never ceased to be a warning
to people who have sold goods on credit.

It should teach the youth of our great, broad land what
glorious possibilities may lie concealed in the rough and
tough bosom of the reluctant present. It shows how steady ·
perseverance and a good appetite will always win in the
end. It teaches us that wealth is not indispensable, and
that if we live as we should, draw out of politics at the
proper time, and die a few days before the public absolutely
demand it, the matter of our birthplace will not be considered.

Still, my birthplace is all right as a birthplace. It was
a good, quiet place in which to be born. All the old neighbors said that Shirley was a very quiet place up to the
time I was born there, and when I took my parents by the
hand and gently led them away in the spring of '53, saying,
" Parents, this is no place for us," it again became quiet.

It is the only birthplace I have, however, and I hope
that all the readers of this sketch will feel perfectly free to
go there any time and visit it and carry their dinner as I
did. Extravagant cordiality and overflowing hospitality
have always kept my birthplace back.

NEVER TALK BACK.

Never talk back! sich things is ripperhensible;
   A feller only "corks" hisse'f that jaws a man that's
      hot;
In a quarrel, ef you'll only keep your mouth shet and
      act sensible,
   The man that does the talkin'll git worsted every
      shot!

Never talk back to a feller that's abusin' you—
   Jest let him carry on, and rip, and cuss and swear;
And when he finds his lyin' and his dammin's jest
      amusin' you,
   You've got him clean kaflummixed, and you want to
      hold him there!

Never talk back, and wake up the whole community,
   And call a man a liar, over law, er Politics,—
You can lift and land him furder and with gracefuller
      impunity
   With one good jolt of silence than a half a dozen
      kicks!

The Gruesome Ballad of Mr. Squincher.

" Ki-yi!" said Mr. Squincher,
   As in contemplative pose,
He stood before the looking-glass
   And burnished up his nose,
And brushed the dandruff from a
     span-
   Spick-splinter suit of clothes,—
" Why, bless you, Mr. Squincher,
   "You're as handsome as a rose!"

"There are some," continued Squin-
    cher,
   As he raised upon his toes
To catch his full reflection,
   And the fascinating bows
That graced his legs,—" I reckon
   There are some folks never knows
How beautiful is human legs
   In pantaloons like those!"

" But ah!" sighed Mr. Squincher,
   As a ghastly phantom 'rose
And leered above his shoulder
   Like the deadliest of foes,—
With fleshless arms and fingers,
   And a skull, with glistening rows
Of teeth that crunched and gritted,—
   " Its my tailor, I suppose !"

*   *   *   *   *   *   *

They found him in the morning—
   So the mystic legend goes—
With the placid face still smiling
   In its statuesque repose ;—
With a lily in his left hand,
   And in his right a rose,
With their fragrance curling upward
   Through a nimbus 'round his nose.

.Anecdotes of Jay Gould.

Facial Neuralgia is what is keeping Jay Gould back this summer and preventing him from making as much money as he would otherwise. With good health and his present methods of doing business Mr. Gould could in a few years be beyond the reach of want, but he is up so much nights with his face that he has to keep one gas-jet burning all the time. Besides he has cabled once to Dr. Brown-Sequard for a neuralgia pill that he thought would relieve the intense pain, and found after he had paid for the cablegram that every druggist in New York kept the Brown-Sequard pill in stock. But when a man is ill he does not care for expense, especially when he controls an Atlantic cable or two.

This neuralgia pill is about the size of a two-year-old colt and pure white. I have been compelled to take several of them myself while suffering from facial neuralgia; for neuralgia does not spare the good, the true or the beautiful.

She comes along and nips the poor yeoman as well as the millionaire who sits in the lap of luxury. Millionaires who flatter themselves that they can evade neuralgia by going and sitting in the lap of luxury make a great mistake.

"And do you find that this large porcelain pill relieves you at all, Mr. Gould?" I asked him during one of these attacks, as he sat in his studio with his face tied up in hot bran.

"No, it does me no good whatever," said the man who likes to take a lame railroad and put it on its feet by issuing more bonds. "It contains a little morphine, which dulls the pain, but there's nothing in the pill to cure the cause. My neuralgia comes from indigestion. My appetite is four sizes too large for a man of my height and every little while I overeat. I then get dangerously ill and stocks become greatly depressed in consequence. I am now in a position where, if I had a constitution that would stand the strain, I could get well off in a few years, but I am not strong enough. Every little change in the weather affects me. I see a red-headed girl on the street and immediately afterwards I see one of these big white pills."

"Are you sure, Mr. Gould?" I asked him with some solicitude, as I bent forward and inhaled the rich fragrance of the carnation in his button-hole, "that you have not taken cold in some way?"

"Possibly I have," he said, as he shrank back in a petulant way, I thought. "Last week I got my feet a little damp while playing the hose on some of my stocks, but I hardly think that was what caused the trouble. I am apt to over-eat, as I said. I am especially fond of fruit, too. When I was a boy I had no trouble, because I always divided my fruit with another boy, of whom I was very fond. I would always divide my fruit into two equal parts, keeping one of these and eating the other myself. Many and many a time when this boy and I went out together and only had one wormy apple between us, I have divided it and given him the worm.

"As a boy, I was taught to believe that half is always better than the hole."

"And are you not afraid that this neuralgia after it has picnicked around among your features may fly to your vitals?"

"Possibly so," said Mr. Gould, snapping the hunting case of his massive silver watch with a loud report, "but I am guarding against this by keeping my pocketbook wrapped up all the time in an old red flannel shirt."

Here Mr. Gould arose and went out of the room for a long time, and I could hear him pacing up and down outside, stopping now and then to peer through the keyhole to

see if I had gone away.   But in each instance he was grat-
ified to find that I had not.   Lest any one should imagine
that I took advantage of his absence to peruse his private
correspondence, I will say here that I did not do so, as his
desk was securely locked.

Mr. Gould's habits are simple and he does not hold his
cane by the middle when he walks.   He wears plain clothes
and his shirts and collars are both made of the same shade.
He says he feels sorry for any one who has to wear a pink
shirt with a blue collar.   Some day he hopes to endow a
home for young men who cannot afford to buy a shirt and a
collar at the same store.

He owes much of his neuralgia to a lack of exercise.
Mr. Gould never takes any exercise at all.   His reason for
this is that he sees no prospect for exercise to advance in
value.   He says he is willing to take anything else but
exercise.

Up to within a very few years Jay Gould has always
slept well at night, owing to regular hours for rising and
retiring and his careful abstinence from tobacco and alco-
hol.   Lately neuralgia has kept him awake a good deal at
night, but prior to that he used to sleep as sweetly and
peacefully as a weasel.

The story circulated some years ago to the effect that a
professional burglar broke into Mr. Gould's room in the
middle of the night and before he could call the police was
robbed of his tools, is not true.   People who have no higher
aim in life than the peddling about of such improbable
yarns would do well to ascertain the truth of these reports
before giving them circulation.

The story that Mr. Gould once killed a steer and presented his hoofs to the poor with the remark that it would
help to keep sole and body together, also turned out to
have no foundation whatever in fact, but was set afloat by
an English wag who was passionately fond of a bit of
pleasantry, don't you know.

Thus it is that a man who has acquired a competence by means of honest toil becomes the target for the barbed shaft of contumely.

Mr. Gould is said to be a good conversationalist, though he prefers to close his eyes and listen to others. Nothing pleases him better than to lure a man on and draw him out and encourage him to turn his mind wrong side out and empty it. He then richly repays this confidence by saying that if it doesn't rain any more we will have a long dry time. The man then goes away inflated with the idea that he has a pointer from Mr. Gould which will materially affect values. A great many men are playing croquet at the poorhouse this summer who owe their prosperity to tips given them by Mr. Gould.

As a fair sample of the way a story about a great man grows and becomes distorted at the same time, one incident will be sufficient. Some years ago, it is said, Mr. Gould bought a general admission ticket to hear Sarah Bernhardt as Camille. Several gentlemen who were sitting near where he stood asked him why he did not take a seat. Instead of answering directly that he could not get one he replied that he did not care for a seat, as he wanted to be near the door when the building fell. Shortly after this he had more seats than he could use. I give this story simply to illustrate how such a thing may be distorted, for upon investigation it was found to have occurred at a Patti concert, and not at a Bernhardt exhibition at all.

Mr. Gould's career, with its attendant success, should teach us two things, at least. One is, that it always pays to do a kind act, for a great deal of his large fortune has been amassed by assisting men like Mr. Field, when they were in a tight place, and taking their depressed stock off their hands while in a shrunken condition. He believes also that the merciful man is merciful to his stock.

He says he owes much of his success in life to economy

and neuralgia. He also loves to relieve distress on Wall street, and is so passionately fond of this as he grows older that he has been known to distress other stock men just for the pleasant thrill it gave him to relieve them.

Jay Gould is also a living illustration of what a young man may do with nothing but his bare hands in America. John L. Sullivan and Gould are both that way. Mr. Gould and Col. Sullivan could go into Siberia to-morrow — little as they are known there — and with a small Gordon press, a quire of bond paper and a pair of three-pennyweight gloves they would soon own Siberia, with a right of way across the rest of Europe and a first mortgage on the Russian throne. As fast as Col. Sullivan knocked out a dynasty Jay could come in and administer on the estate. This would be a powerful combination. It would afford us an opportunity also to get some of those Russian hay-fever names and chilblains by red message. Mr. Gould would get a good deal of money out of the transaction and Sullivan would get ozone.

A Fall-Crick View of the Earthquake

I kin hump my back and take the rain,
  And I don't keer how she pours;
I kin keep kindo' ca'm in a thunder storm,
  No matter how loud she roars;
I haint much skeered o' the lightnin',
  Ner I haint sich awful shakes
Afeared o' *cyclones*—but I don't want none
  O' yer dad-burned old *earth-* quakes!

As long as my legs keeps stiddy,
  And long as my head keeps plum,
And the buildin' stays in the front lot,
  I still kin whistle, *some!*
But about the time the old clock
  Flops off'n the mantel-shelf,
And the bureau skoots fer the kitchen,
  I'm a-goin' to skoot, myself!

Plague-take! ef you keep me stabled
  While any earthquakes is around!—
I'm jist like the stock,—I'll beller,

And break fer the open ground!
And I 'low you'd be as nervous,
    And in jist about my fix,
When yer whole farm slides from inunder you,
    And on'y the mor'gage sticks!

Now cars haint a-goin' to kill you
    Ef you don't drive 'crost the track;
Crediters never'll jerk you up
    Ef you go and pay 'em back;
You kin stand all moral and mundane storms
    Ef you'll on'y jist behave—
But a' EARTHQUAKE:—well, ef it wanted you,
    It 'ud husk you out o' yer grave!

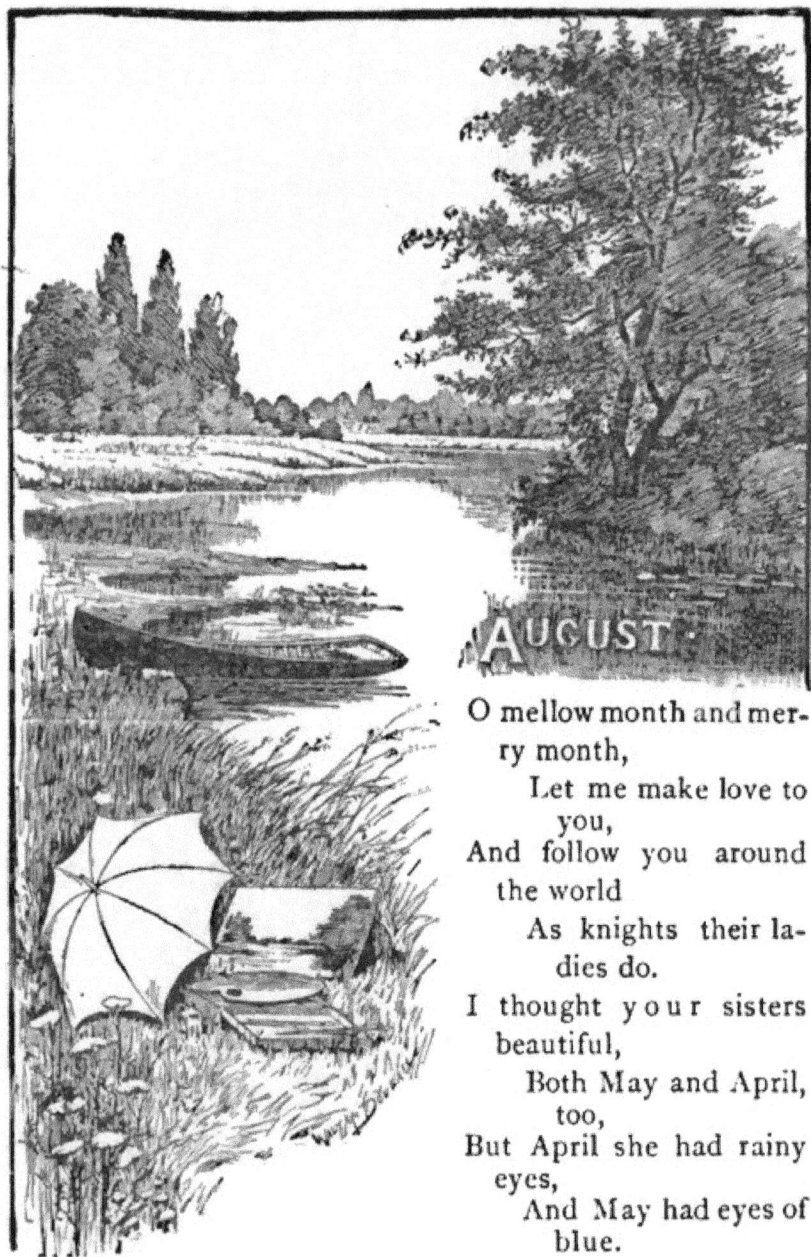

AUGUST

O mellow month and mer-
ry month,
 Let me make love to
 you,
And follow you around
the world
 As knights their la-
 dies do.
I thought your sisters
beautiful,
 Both May and April,
 too,
But April she had rainy
eyes,
 And May had eyes of
 blue.

And June—I liked the singing
　　Of her lips, and liked her smile—
But all her songs were promises
　　Of something, *after while*;
And July's face—the lights and shade
　　That may not long beguile,
With alternations o' er the wheat,
　　The dreamer at the stile.

But you!—ah, you are tropical,
　　Your beauty is so rare:
Your eyes are clearer, deeper eyes
　　Than any, anywhere;
Mysterious, imperious,
　　Deliriously fair,
O listless Andalusian maid,
　　With bangles in your hair!

Julius Caesar in Town.

THE PLAY of "Julius Cæsar," which has been at the Academy of Music this week, has made a great hit. Messrs. Booth and Barrett very wisely decided that if it succeeded here it would do well anywhere. If the people of New York like a play and say so, it is almost sure to go elsewhere. Judging by this test the play of "Julius Cæsar" has a glowing future ahead of it. It was written by Gentlemen Shakespeare, Bacon and Donnelly, who collaborated together on it. Shakespeare did the lines and plot, Bacon furnished the cipher and Donnelly called attention to it through the papers.

The scene of "Julius Cæsar" is laid in Rome just before the railroad was completed to that place. In order to understand the play itself we must glance briefly at the leading characters which are introduced and upon whom its success largely depends.

Julius Cæsar first attracted attention through the Roman papers by calling the attention of the medical faculty to the now justly celebrated Cæsarian operation. Taking advantage of the advertisement thus attained, he soon rose to prominence and flourished considerably from 100 to 44 B.C., when a committee of representative citizens and property-owners of Rome called upon him and on behalf of the people begged leave to assassinate him as a mark of esteem. He was stabbed twenty-three times between Pompey's Pillar and eleven o'clock, many of which were mortal. This account of the assassination is taken from a local paper and is graphic, succinct and lacks the sensational elements so common and so lamentable in our own time. Cæsar was the implacable foe of the aristocracy and refused to wear a plug hat up to the day of his death. Sulla once said, before Cæsar had made much of a showing, that some day this young man would be the ruin of the aristocracy, and twenty years afterwards when Cæsar sacked, assassinated and holocausted a whole theological seminary for saying "eyether" and "nyether," the old settlers recalled what Sulla had said.

Cæsar continued to eat pie with a knife and in many other ways to endear himself to the masses until 68 B.C., when he ran for Quæstor. Afterward he was Ædile, during the term of which office he sought to introduce a number of new games and to extend the limit on some of the older ones. From this to the Senate was but a step. In the Senate he was known as a good Speaker, but ambitious, and liable to turn up during a close vote when his enemies thought he was at home doing his chores. This made him at times odious to those who opposed him, and when he defended Cataline and offered to go on his bond, Cæsar came near being condemned to death himself.

In 62 B. C. he went to Spain as Proprætor, intending to write a book about the Spanish people and their customs as soon as he got back, but he was so busy on his return that he did not have time to do so.

Cæsar was a powerful man with the people, and while in the Senate worked hard for his constituents, while other Senators were having their photographs taken. He went into the army when the war broke out, and after killing a great many people against whom he certainly could not have had anything personal, he returned, headed by the Rome Silver Cornet Band and leading a procession over two miles in length. It was at this time that he was tendered a crown just as he was passing the City Hall, but thrice he refused it. After each refusal the people applauded and encored him till he had to refuse it again. It is at about this time the play opens. Cæsar has just arrived on a speckled courser and dismounted outside the town. He comes in at the head of the procession with the understanding that the crown is to be offered him just as he crosses over to the Court-House.

Here Cassius and Brutus meet, and Cassius tries to make a Mugwump of Brutus, so that they can organize a new movement. Mr. Edwin Booth takes the character of Brutus and Mr. Lawrence Barrett takes that of Cassius. I would not want to take the character of Cassius myself, even if I had run short of character and needed some very much indeed, but Mr. Barrett takes it and does it first-rate. Mr. Booth also plays Brutus so that old settlers here say it seems almost like having Brutus here among us again.

Brutus was a Roman republican with strong tariff tendencies. He was a good extemporaneous after-dinner speaker and a warm personal friend of Cæsar, though differing from him politically. In assassinating Cæsar, Brutus used to say afterwards he did not feel the slightest personal animosity, but did it entirely for the good of the party. That is one thing I like about politics—you can cut out a man's vitals and hang them on the Christmas tree and drag

the fair name of his wife or mother around through the
sewers for six weeks before election, and so long as it is
done for the good of the party it is all right.

So when Brutus is authorized by the caucus to assassinate Cæsar he feels that, like being President of the
United States, it is a disagreeable job; but if the good of
the party seems really to demand it he will do it, though he
wishes it distinctly understood that personally he hasn't got
a thing against Cæsar.

In act 4 Brutus sits up late reading a story by E. P. Roe,
and just as he is in the most exciting part of it the ghost of
the assassinated Cæsar appears and states that it will meet
him with hard gloves at Philippi. Brutus looks bored and
says that he is not in condition, but the ghost leaves it that
way and Brutus looks still more bored till the ghost goes
out through a white oak door without opening it.

At Philippi, Brutus sees that there is no hope of police
interference, and so before time is called he inserts his
sword into his being and dies while the polite American
audience puts on its overcoat and goes out, looking over its
shoulder to see that Brutus does not take advantage of this
moment, while the people are going away, to resuscitate
himself.

The play is thoroughly enjoyable all the way through,
especially Cæsar's funeral. The idea of introducing a
funeral and engaging Mark Antony to deliver the eulogy,
with the understanding that he was to have his traveling
expenses paid and the privilege of selling the sermon to a
syndicate, shows genius on the part of the joint authors.
All the way through the play is good, but sad. There is
no divertisement or tank in it, but the funeral more than
makes up for all that.

Where Portia begs Brutus, before the assassination, to
tell her all and let her in on the ground floor, and asks
what the matter is, and he claims that it is malaria, and she

still insists and asks, "Dwell I but in the suburbs of your good pleasure?" and he states, "You are my true and honorable wife, as dear to me as are the ruddy drops that visit my sad heart," I forgot myself and wept my new plug hat two-thirds full. It is as good as anything there is in Josh Whitcomb's play.

Booth and Barrett have the making of good actors in them. I met both of these gentlemen in Wyoming some

years ago. We met by accident. They were going to California and I was coming back. By some oversight we had both selected the same track, and we were thrown together. I do not know whether they will recall my face or not. I was riding on the sleeper truck at the time of the accident. I always take a sleeper and always did. I rode on the truck because I didn't want to ride inside the car and have to associate with a wealthy porter who looked down upon me. I am the man who was found down the creek the next day gathering wild ferns and murmuring, "Where am I?"

The play of "Julius Cæsar" is one which brings out the meanness and magnetism of Cassius, and emphasizes the mistaken patriotism of Brutus. It is full of pathos, duplicity, assassination, treachery, erroneous loyalty, suicide, hypocrisy, and all the intrigue, jealousy, cowardice and deviltry which characterized the politics of fifty years B.C., but which now, thanks to the enlightenment and refinement which twenty centuries have brought, are known no more forever. Let us not forget, as we enter upon the year 1888, that it is a Presidential year, and that all acrimony will be buried under the dew and the daisies, and that no matter how high party spirit may run, there will be no personal enmity.

His First Womern.

I buried my first womern
  In the spring; and in the
    fall
I was married to my second,
  And haint settled yit at
    all!—
Fer I'm allus thinkin'—
    thinkin'
  Of the first one's peaceful
    ways,
A-bilin' soap and singin'
  Of the Lord's amazin'
    grace.

And I'm thinkin' of her, constant,
   Dyin' carpet-chain and stuff,
And a-makin' up rag-carpets,
   When the floor was good enough!
And I mind her he'p a-feedin',
   And I recollect her now
A-drappin' corn, and keepin'
   Clos't behind me and the plow!

And I'm allus thinkin' of her
   Reddin' up around the house;
Er cookin' fer the farm-hands;
   Er a-drivin' up the cows.—
And there she lays out yender
   By the lower medder-fence,
Where the cows was barely grazin',
   And they're usin' ever sence.

And when I look acrost there—
   Say its when the clover's ripe,
And I'm settin', in the evenin',
   On the porch here, with my pipe,
And the *other'n* hollers "Henry!"—
   W'y they ain't no sadder thing
Than to think of my first womern
   And her funeral last spring
     Was a year ago.

This Man Jones.

This man Jones was what you'd call
A feller 'at had no sand at all;
Kindo consumpted, and undersize,
And saller-complected, with big sad eyes,
And a kind-of-a-sort-of-a-hang-dog style,
And a sneakin' kind-of-a-half-way smile
That kindo give him away to us
As a preacher, maybe, or somepin' wuss.

Didn't take with the gang—well, no—
But still we managed to *use* him, though,—
Coddin' the gilley along the rout'
And drivin' the stakes that he pulled out;—
For I was one of the bosses then
And of course stood in with the canvassmen—'
And the way we put up jobs, you know,
On this man Jones jes' beat the show!

Used to rattle him scandalous,
And keep the feller a-dodgin' us,
And a-shyin' round jes' skeered to death,
And a-feered to whimper above his breath;
Give him a cussin', and then a kick,
And then a kind-of-a back-hand lick—
Jes' for the fun of seein' him climb
Around with a head on half the time.

But what was the curioust thing to me,
Was along o' the party—let me see,—
Who was our "Lion Queen" last year?—
Mamzelle Zanty, er De La Pierre?—
Well, no matter!—a stunnin' mass,
With a red-ripe lip, and a long eye-lash,
And a figger sich as the angels owns—
And one too many for this man Jones!

He'd always wake in the afternoon
As the band waltzed in on "the lion tune,"
And there, from the time that she'd go in,
Till she'd back out of the cage agin,
He'd stand, shaky and limber-kneed—
'Specially when she come to "feed
The beast raw meat with her naked hand"—
And all that business, you understand.

And it *was* resky in that den—
For I think she juggled three cubs then,
And a big "green" lion 'at used to smash
Collar-bones fer old Frank Nash;
And I reckon now she haint forgot
The afternoon old "Nero" sot
His paws on her:—but as fer me,
Its a sort-of-a-mixed-up mystery.

Kindo' remember an awful roar,
And see her back fer the bolted door—
See the cage rock—heerd her call
"God have mercy!" and that was all—
For thar haint no livin' man can tell
What it's like when a thousand yell
In female tones, and a thousand more
Howl in bass till their throats is sore!

But the keeper said as they dragged her out,
They heerd some feller laugh and shout:
"Save her! Quick! I've got the cuss!"
......And yit she waked and smiled on us!
And we daren't *flinch*—fer the doctor said,
Seein' as this man Jones was dead,
Better to jes' not let her know
Nothin' o' that fer a week or so.

How to Hunt the Fox.

THE joyous season for hunting is again upon us, and with the gentle fall of the autumn leaf and the sough of the scented breezes about the gnarled and naked limbs of the wailing trees — the huntsman comes with his hark and his halloo and hurrah, boys, the swift rush of the chase, the thrilling scamper 'cross country, the mad dash through the Long Islander's pumpkin patch—also the mad dash, dash, dash of the farmer, the low moan of the disabled and frozen-toed hen as the whooping horsemen run her down; the wild shriek of the children, the low melancholy wail of the frightened shoat as he flees away to the straw pile, the quick yet muffled plunk of the frozen tomato and the dull scrunch of the seed cucumber.

The huntsman now takes the flannels off his fox, rubs his stiffened limbs with gargling oil, ties a bunch of firecrackers to his tail and runs him around the barn a few times to see if he is in good order.

The foxhound is a cross of the bloodhound, the grayhound, the bulldog and the chump. When you step on his tail he is said to be in full cry. The foxhound obtains from his ancestors on the bloodhound side of the house his keen scent, which enables him while in full cry 'cross country to pause and hunt for chipmunks. He also obtains from the bloodhound branch of his family a wild yearning to star in an "Uncle Tom" company, and watch little Eva meander up the flume at two dollars per week. From the grayhound he gets his most miraculous speed, which enables him to attain a rate of velocity so great that he is unable to halt during the excitement of the chase, frequently running so far during the day that it takes him a week to get back, when, of course, all interest has died out. From the bulldog the foxhound obtains his great tenacity of purpose, his deep-seated convictions, his quick perceptions, his love of home and his clinging nature. From the chump the foxhound gets his high intellectuality and that mental power which enables him to distinguish almost at a glance the salient points of difference between a two-year-old steer and a two-dollar bill.

The foxhound is about two feet in height, and 120 of them would be considered an ample number for a quiet little fox hunt. Some hunters think this number inadequate, but unless the fox be unusually skittish and crawl under the barn, 120 foxhounds ought to be enough. The trouble generally is that hunters make too much noise, thus scaring the fox so that he tries to get away from them. This necessitates hard riding and great activity on the part of the whippers-in. Frightening a fox almost always results in sending him out of the road and compelling horsemen to stop in order to take down a panel of fence every little while that they may follow the animal, and before you can get the fence put up again the owner is on the ground, and after you have made change with him and mounted again the fox may be nine miles away. Try by all means to keep your fox in the road!

It makes a great difference what kind of fox you use, however. I once had a fox on my Pumpkin Butte estates that lasted me three years, and I never knew him to shy or turn out of the road for anything but a loaded team. He was the best fox for hunting purposes that I ever had. Every spring I would sprinkle him with Scotch snuff and put him away in the bureau till fall. He would then come out bright and chipper. He was always ready to enter into the chase with all the chic and embonpoint of a regular Kenosha, and nothing pleased him better than to be about eight miles in advance of my thoroughbred pack in full cry, scampering 'cross country, while stretching back a few miles behind the dogs followed a pale young man and his fiancier, each riding a horse that had sat down too hard on its tail some time and driven it into his system about six joints.

Some hunters, who are madly and passionately devoted to the sport, leap their horses over fences, moats, donjon keeps, hedges and current bushes with utter sang froid and the wild, unfettered toot ongsomble of a brass band. It is one of the most spirited and touchful of sights to see a young fox-hunter going home through the gloaming with a full cry in one hand and his pancreas in the other.

Some like to be in at the death, as it is called, and it is certainly a laudable ambition. To see 120 dogs hold out against a ferocious fox weighing nine pounds; to watch the brave little band of dogs and whippers-in and horses with sawed-off tails, making up in heroism what they lack in numbers, succeeding at last in ridding the country of the ferocious brute which has long been the acknowledged foe of the human race, is indeed a fine sight.

We are too apt to regard fox-hunting merely as a relaxa-tion, a source of pleasure, and the result of a desire to do the way people do in the novels which we steal from Eng-lish authors: but this is not all. To successfully hunt a fox, to jump fences 'cross country like an unruly steer, is no child's play. To ride all day on a very hot and restless saddle, trying to lope while your horse is trotting, giving your friends a good view of the country between yourself and your horse, then leaping stone walls, breaking your col-lar-bone in four places, pulling out one eye and leaving it hanging on a plum tree, or going home at night with your transverse colon wrapped around the pommel of your sad-dle and your liver in an old newspaper, requires the great-est courage.

Too much stress cannot be placed upon the costume worn while fox-hunting, and in fact, that is, after all, the life and soul of the chase. For ladies, nothing looks better than a close-fitting jacket, sewed together with thread of the same shade and a skirt. Neat-fitting cavalry boots and a plug hat complete the costume. Then, with a hue in one hand and a cry in the other, she is prepared to mount. Lead the horse up to a stone wall or a freight car and spring lightly into the saddle with a glad cry. A freight car is the best thing from which to mount a horse, but it is too unwieldy and frequently delays the chase. For this reason, too, much luggage should not be carried on a fox-hunt. Some gentlemen carry a change of canes, neatly con-cealed in a shawl strap, but even this may be dispensed with.

For gentlemen, a dark, four-button cutaway coat, with neat, loose-fitting, white panties, will generally scare a fox into convulsions, so that he may be easily killed with a club. A short-waisted plug hat may be worn also, in order to distinguish the hunter from the whipper-in, who wears a baseball cap. The only fox-hunting I have ever done was on board an impetuous, tough-bitted, fore-and-aft horse that had emotional insanity. I was dressed in a swallow-tail coat, waistcoat of Scotch plaid Turkish toweling, and a pair of close-fitting breeches of etiquette tucked into my boot-tops. As I was away from home at the time and could not reach my own steed I was obliged to mount a spirited steed with high, intellectual hips, one white eye and a big red nostril that you could set a Shanghai hen in. This horse, as soon as the pack broke into full cry, climbed over a fence that had wrought-iron briers on it, lit in a corn field, stabbed his hind leg through a sere and yellow pumpkin, which he wore the rest of the day, with seven yards of pumpkin vine streaming out behind, and away

we dashed 'cross country. I remained mounted not because I enjoyed it, for I did not, but because I dreaded to dismount. I hated to get off in pieces. If I can't get off a horse's back as a whole, I would rather adhere to the horse. I will adhere that I did so.

We did not see the fox, but we saw almost everything else. I remember, among other things, of riding through a hothouse and how I enjoyed it. A morning scamper through a conservatory when the syringas and Jonquils

and Jack roses lie cuddled up together in their little beds, is a thing to remember and look back to and pay for. To stand knee-deep in glass and gladiolas, to smell the mashed and mussed up mignonette and the last fragrant sigh of the scrunched heliotrope beneath the hoof of your horse, while far away the deep-mouthed baying of the hoarse hounds, hotly hugging the reeking trail of the anise-seed bag, calls on the gorgeously caparisoned hills to give back their merry music or fork it over to other answering hills, is joy to the huntsman's heart.

On, on I rode with my unconfined locks streaming behind me in the autumn wind. On and still on I sped, the big, bright pumpkin slipping up and down the gambrel of my spirited horse at every jump. On and ever on we went, shedding terror and pumpkin seeds along our glittering track till my proud steed ran his leg in a gopher hole and fell over one of those machines that they put on a high-headed steer to keep him from jumping fences. As the horse fell, the necklace of this hickory poke flew up and adjusted itself around my throat. In an instant my steed was on his feet again, and gayly we went forward while the prong of this barbarous appliance, ever and anon plowed into a brand new culvert or rooted up a clover field. Every time it ran into an orchard or a cemetery it would jar my neck and knock me silly. But I could see with joy that it reduced the speed of my horse. At last as the sun went down, reluctantly, it seemed to me, for he knew that he would never see such riding again, my ill-spent horse fell with a hollow moan, curled up, gave a spasmodic quiver with his little, nerveless, sawed-off tail and died.

The other huntsmen succeeded in treeing the anise-seed bag at sundown, in time to catch the 6 o'clock train home.

Fox-hunting is one of the most thrilling pastimes of which I know, and for young men whose parents have amassed large sums of money in the intellectual pursuit of hides and tallow, the meet, the chase, the scamper, the full cry, the cover, the stellated fracture, the yelp of the pack, the yip, the yell of triumph, the confusion, the whoop, the holla, the hallos, the hurrah, the abrasion, the snort of the hunter, the concussion, the sward, the open, the earth stopper, the strangulated hernia, the glad cry of the hound as he brings home the quivering seat of the peasant's pantaloons, the yelp of joy as he lays at his master's feet, the strawberry mark of the rustic, all, all are exhilarating to the sons of our American nobility.

Fox-hunting combines the danger and the wild, tumultuous joy of the skating-rink, the toboggan slide, the mush-and-milk sociable and the straw ride.

With a good horse, an air cushion, a reliable earth-stopper and an anise-seed bag, a man must indeed be thoroughly blase who can not enjoy a scamper across country, over the Pennsylvania wold, the New Jersey mere, the Connecticut moor, the Indiana glade, the Missouri brake, the Michigan mead, the American tarn, the fen, the gulch, the buffalo wallow, the cranberry marsh, the glen, the draw, the canyon, the ravine, the forks, the bottom or the settlement.

For the young American nobleman whose ducal father made his money by inventing a fluent pill, or who gained his great wealth through relieving humanity by means of a lung pad, a liver pad, a kidney pad or a foot pad, fox-hunting is first-rate.

The Boy-Friend.

LARENCE, my boy-friend,
   hale and strong,
  O, he is as jolly as he is
   young;
And all of the laughs of the
   lyre belong
  To the boy all unsung:

So I want to sing somthing in
   his behalf—
To clang some chords, for the good it is
To know he is near, and to have the laugh
  Of that wholesome voice of his.

I want to tell him in gentler ways
  Than prose may do, that the arms of rhyme,
Warm and tender with tuneful praise,
  Are about him all the time.

I want him to know that the quietest nights
  We have passed together are yet with me,
Roistering over the old delights
  That were born of his company.

I want him to know how my soul esteems
  The fairy stories of Anderson,
And the glad translations of all the themes
  Of the hearts of boyish men.

Want him to know that my fancy flows,
  With the lilt of a dear old-fashioned tune,
Through "Lewis Carroll's" poemly prose,
  And the tale of "The Bold Dragoon."

O, this is the Prince that I would sing—
  Would drape and garnish in velvet line,
Since courtlier far than any king
  Is this brave boy-friend of mine!

A Letter of Acceptance.

The secretary of the Ashfield Farmers' Club, of Ashfield, Mass., Mr. E. D. Church, informs me by United States mail that upon receipt of my favorable reply I will become an honorary member of that club, along with George William Curtis, Prof. Norton, Prof. Stanley Hall, of Harvard, and other wet-browed toilers in the catnip-infested domain of Agriculture.

I take this method of thanking the Ashfield Farmers' Club, through its secretary, for the honor thus all so unworthily bestowed, and joyfully accept the honorary membership, with the understanding, however, that during the County Fair the solemn duty of delivering the annual address from the judges' stand, in tones that will not only ring along down the corridors of time, but go thundering three times around a half-mile track and be heard above the rhythmic plunk of the hired man who is trying to ascertain, by means of a large mawl and a thumping machine, how hard he can strike, shall fall upon Mr. Curtis or other honorary members of the club. I have a voice that does very well to express endearment, or other subdued emotions, but it is not effective at a County Fair. Spectators see the wonderful play of my features, but they only hear the low refrain of the haughty Clydesdale steed, who has a neighsal voice and wears his tail in a Grecian coil. I received $150 once for addressing a race-track one mile in length on "The Use and Abuse of Ensilage as a Narcotic." I made the gestures, but the sentiments were those of the four-ton Percheron charger, Little Medicine, dam Eloquent.

I spoke under a low shed and rather adverse circumstances. In talking with the committee afterwards, as I wrapped up my gestures and put them back in the shawl

strap, I said that I felt almost ashamed to receive such a price for the sentiments of others, but they said that was all right. No one expected to hear an Agricultural Address.

They claimed that it was most generally purely spectacular, and so they regarded my speech as a great success. I used the same gestures afterwards in speaking of "The Great Falling Off Among Bare-Back Riders in the Circuses of the Present Day."

I would also like to be excused from any duties as a judge of curly-faced stock or as an umpire of ornamental needlework. After a person has had a fountain pen kicked endwise through his chest by the animal to which he has awarded the prize, and later on has his features worked up into a giblet pie by the owner of the animal to whom he did not award the prize, he does not ask for public recognition at the hands of his fellow-citizens. It is the same in the matter of ornamental needlework and gaudy quilts, which goad a man to drink and death. While I am proud to belong to a farmers' club and "change works" with a hearty, whole-souled ploughman like George William Curtis, I hope that at all County Fairs or other intellectual hand-to-hand contests between outdoor orators and other domestic animals, I may be excused, and that when judges of inflamed slumber robes and restless tidies, which roll up and fall over the floor or adhere to the backs of innocent people; or stiff, hard Doric pillow-shams which do not in any way enhance the joys of sleep; or beautiful, pale-blue satin pincushions, which it would be wicked to put a pin in and which will therefore ever and forevermore mock the man who really wants a pin, just as a beautiful match-safe stands idly through the long vigils of the night, year after year, only to laugh at the man who staggers towards it and falls up against it and finds it empty; or like the glorious inkstand which is so pretty and so fragile that it stands around with its hands in its pockets acquiring dust and dead flies for centuries, so that when you are in a hurry you stick your pen into a small chamber of horrors—I say when the judges are selected for this department I would rather have my name omitted from the panel, as I have formed or expressed an opinion and have reasonable doubts and conscientious scruples which it would require testimony to remove, and I am not qualified anyway, and I have been already placed in jeopardy once, and that is enough.

Mr. Church writes that the club has taken up, discussed and settled all points of importance bearing upon Agriculture, from the tariff up to the question of whether or not turpentine poured in a cow's ear ameliorates the pangs of hollow horn. He desires suggestions and questions for discussion. That shows the club to be thoroughly alive. It will soon be Spring, and we cannot then discuss these matters. New responsibilities will be added day by day in the way of stock, and we will have to think of names for them. Would it not be well before the time comes for active farm work to think out a long list of names before the little strangers arrive? Nothing serves to lower us in the estimation of our fellow-farmers or the world more than the frequent altercations between owners and their hired help over what name they shall give a weary, wobbly calf who has just entered the great arena of life, full of hopes and aspirations, perhaps, but otherwise absolutely empty. Let us consider this before Spring fairly opens, so that we may be prepared for anything of this kind.

One more point may properly come before the club at its next meeting, and I mention it here because I may be so busy at Washington looking after our other interests that I cannot get to the club meeting. I refer to the evident change in climate here from year to year, and its effect upon seeds purchased of florists and seedsmen generally.

Twenty years ago you could plant a seed according to directions and it would produce a plant which seemed to resemble in a general way the picture on the outside of the package. Now, under the fluctuating influences of irresponsible isotherms, phlegmatic Springs, rare June weather and overdone weather in August, I find it almost impossible to produce a plant or vegetable which in any way resembles its portrait. Is it my fault or the fault of the climate? I wish the club would take hold of this at its next regular meeting. I first noticed the·change in the summer of '72, I think. I purchased a small package of early Scotch plaid curled kale with a beautiful picture on the outside. It was as good a picture of Scotch kale as I ever saw. I could imagine how gay and light-hearted it was the day it went up to the studio and had its picture taken for this purpose. A short editorial paragraph under the picture stated that I should plant in quick, rich soil, in rows four inches apart, to a depth of one inch, cover lightly and then roll. I did so. No farmer of my years enjoys rolling better than I do.

In a few weeks the kale came up but turned out to be a canard. I then waited two weeks more and other forms of vegetation made their appearance. None of them were kale. A small delegation of bugs which deal mostly with kale came into the garden one day, looked at the picture on the discarded paper, then examined what had crawled out through the ground and went away. I began to fear then that climatic influences had been at work on the seeds, but I had not fully given up all hope.

At first the plants seemed to waver and hesitate over whether they had better be wild parsnips or Lima beans. Then I concluded that they had decided to be foliage plants or rhubarb. But they did not try to live up to their portraits. Pretty soon I discovered that they had no bugs which seemed to go with them, and then I knew they were weeds. Things that are good to eat always have bugs and worms on them, while tansy and castor-oil go through life unmolested.

I ordered a new style of gladiola eight years ago of a man who had his portrait in the bow of his seed catalogue. If he succeeds no better in resembling his portrait than his gladiolas did in resembling theirs, he must be a human onion whose presence may easily be detected at a great distance.

Last year I planted the seeds of a watermelon which I bought of a New York seedsman who writes war articles winters and sells garden seeds in the Spring. The portrait of this watermelon would tempt most any man to climb a nine-rail fence in the dead of night and forget all else in order to drown his better nature and his nose in its cool bosom. People came for miles to look at the picture of this melon and went away with a pleasant taste in their mouths.

The plants were a little sluggish, though I planted in hills far apart each way in a rich warm loam enriched by everything that could make a sincere watermelon get up and hump itself. The melons were to be very large indeed, with a centre like a rose. According to the picture, these melons generally grew so large and plenty that most everybody had to put side-boards on the garden fence to keep them from falling over into other farms and annoying people who had all the melons they needed. I fought squash bugs, cut worms, Hessian flies, chinch bugs, curculio, mange, pip, drought, dropsy, caterpillars and contumely till the latter part of August, when a friend from India came to visit me. I decided to cut a watermelon in honor of his arrival. When the proper moment had arrived and the dinner had progressed till the point of fruit, the tropical depths of my garden gave up their season's wealth in the shape of a low-browed citron about as large and succulent as a hot ball.

I have had other similar experiences, and I think we ought to do something about it if we can. I have planted the seed of the morning glory and the moon flower and dreamed at night that my home looked like a florist's advertisement, but when leafy June came a bunch of Norway oats and a hill of corn were trying to climb the strings nailed up for the use of my non-resident vines. I have planted with song and laughter the seeds of the ostensible pansy and carnation, only in tears to reap the bachelor's button and the glistening foliage of the sorghum plant. I have planted in faith and a deep, warm soil, with pleasing hope in my heart and a dark-red picture on the outside of the package, only to harvest the low, vulgar jimson weed and the night-blooming bull thistle.

Does the mean temperature or the average rainfall have anything to do with it? If statistics are working these changes they ought to be stopped. For my own part, however, I am led to believe that our seedsmen put so much money into their catalogues that they do not have anything left to use in the purchase of seeds. Good religion and very fair cookies may be produced without the aid of caraway seed, but you cannot gather nice, fresh train figs of thistles or expect much of a seedsman whose plants make no effort whatever to resemble their pictures.

Hoping that you will examine into this matter, and that the club will always hereafter look carefully in this column for its farm information, I remain, in a sitting posture, yours truly. BILL NYE.

"YOU IN THE HAMMOCK; AND I, NEAR BY."

In the Afternoon.

You in the hammock; and I, near by,
  Was trying to read, and to swing you, too;
And the green of the sward was so kind to the eye,
  And the shade of the maples so cool and blue,
  That often I looked from the book to you
To say as much, with a sigh.

You in the hammock. The book we'd brought
  From the parlor—to read in the open air,—
Something of love and of Launcelot
  And Guinevere, I believe, was there—
  But the afternoon, it was far more fair
Than the poem was, I thought.

You in the hammock; and on and on
  I droned and droned through the rythmic stuff—
But with always a half of my vision gone
  Over the top of the page—enough
  To caressingly gaze at you, swathed in the fluff
Of your hair and your odorous lawn.

You in the hammock—And that was a year—
  Fully a year ago, I guess!—
And what do we care for their Guinevere
  And her Launcelot and their lordliness!—
  You in the hammock still, and—Yes—
Kiss me again, my dear!

The Rise and Fall of William Johnson.

(A CHRISTMAS STORY.)

WM JOHNSON.

IT HAS always been one of my pet notions that on Christmas day we ought not to remember those only who may be related to us and those who are prosperous, but, that we should, while remembering them, forget not the unfortunate who are dead to all the world but themselves and who suffer in prison walls, not alone for their own crimes, perhaps, but for the crimes of their parents and their grandparents before them. Few of the prosperous and happy pause today to think of the convict whose days are all alike and whose nights are filled with bitterness.

At the risk of being dull and prosy, I am going to tell a story that is not especially humorous or pathetic, but merely true. Every Christmas I try to tell a true story. I do not want the day to go by without some sort of recognition by which to distinguish it from other days, and so I celebrate it in that way.

This is the story of William Johnson, a Swede, who went to Wyoming Territory, perhaps fifteen years ago, to seek his fortune among strangers, and who, without even a knowledge of the English language, began in his patient way to work at whatever his hands found to do. He was a plain, long-legged man, with downcast eyes and nose.

There was some surprise expressed all around when he was charged one day by Jake Feinn with feloniously taking, stealing, carrying away and driving away one team of horses, the property of the affiant, and of the value of $200, contrary to the statutes in such case made and provided, and against the peace and dignity of the Territory of Wyoming.

Everybody laughed at the idea of Jake Feinn owning a team worth $200, and, as he was also a chronic litigator, it was generally conceded that Johnson would be discharged. But his misfortunes seemed to swoop down on him from the very first moment. At the preliminary examination Johnson acted like a man who was dazed. He couldn't talk or understand English very well. He failed to get a lawyer. He pleaded guilty, not knowning what it meant, and was permitted to take it back. He had no witnesses, and the Court was in something of a hurry as it had to prepare a speech that afternoon to be delivered in the evening on the "Beauties of Eternal Justice," and so it was adjudged that in default of $500 bail the said William Johnson be committed to the County Jail of Albany County in said Territory, there to await the action of the Grand Jury for the succeeding term of the District Court for the Second Judicial District of Wyoming.

Meekly and silently William Johnson left the warm and stimulating Indian summer air of October to enter the dark and undesirable den of a felon. Patiently he accepted the heart-breaking destiny which seemed really to belong to someone else. He put in his days studying an English primer all the forenoon and doing housework around the jail kitchen in the afternoon.

He was a very tall man and a very awkward man, with large, intellectual joints and a sad face. When he got so that he could read a little I went in to hear him one day.

He stood up like an exaggerated schoolboy, and while he bored holes in the page of his primer with a long and corneous forefinger he read that little poem:

Pray tell me, bird, what you can see
Up in the top of that tall tree?
Have you no fear that some rude boy
May come and mar your peace and joy?
   *   *   *   *   *   *
Oh, no, my child, I fear no harm,
While with my song I thus can charm.
My mate is here, my youngsters, too,
And here we sit and sing to you.

Finally, the regular term of the District Court opened. Men who had come for a long distance to vaunt their ignorance and other qualifications as jurors could be seen on the streets. Here and there you could see the familiar faces of those who had served as jurors for years and yet had never lost a case. Wealthy delinquents began to subpœna large detachments of witnesses at the expense of the county, and the poor petty larceny people in the jail began to wonder why their witnesses didn't show up. Slowly the wheels of Justice began to revolve. Ever and anon could be heard the strident notes which came from the room where the counsel for the defense was filing his objections, while now and then the ear was startled with the low quash of the indictment.

Finally the case of the Territory against William Johnson was called.

" Mr. Johnson," asked Judge Blair, " have you counsel?"

The defendant said he had not.

" Are you able to employ counsel ? "

He evidently wasn't able to employ counsel twenty minutes, even if it could be had at a dollar a day.

" Do you wish to have the Court appoint counsel for you ? "

He saw no other way, so he said yes.

Where criminals are too poor to employ counsel the Court selects a poor but honest young lawyer, who practices on the defendant. I was appointed that way myself once to defend a man who swears he will kill me as soon as he gets out of the penitentiary.

Wm. Johnson was peculiarly unfortunate in the election of his counsel. The man who was appointed to defend him was a very much overestimated young man who started the movement himself. He was courageous, however, and perfectly willing to wade in where angels would naturally hang back. His brain would not have soiled the finest fabric, but his egotism had a biceps muscle on it like a loaf of Vienna bread. He was the kind of young man who loves to go and see the drama and explain it along about five minutes in advance of the company in a loud, trenchant voice.

He defended William Johnson. Thus in the prime of life, hardly understanding a word of the trial, stunned, helpless, alone, the latter began upon his term of five years in the penitentiary. His patient, gentle face impressed me as it did others, and his very helplessness thus became his greatest help.

It is not egotism which prompts me to tell here of what followed. It was but natural that I should go to Judge Blair, who, besides being the most popular Judge in the West, had, as I knew, a kind heart. He agreed with me that Johnson's side of the case had not been properly presented and that the jury had grave doubts about the horses having been worth enough to constitute a felony even if Johnson had unlawfully taken them. Other lawyers said that at the worst it was a civil offense, or trover, or trespass, or wilful negligence, or embezzlement, or conversion, but that the remedy was by civil process. One lawyer said it was an outrage, and Charlie Bramel said that if Johnson would put up $50 he would agree to jerk him out of the jug on a writ of habeas corpus before dinner.

Seeing how the sentiment ran, I resolved to start a petition for Johnson's pardon. I got the signatures of the Court, the court officers, the jury and the leading men of business in the county. Just as I was about to take it to Gov. Thayer, there was an incident at the penitentiary. Wm. Johnson had won the hearts of the Warden and the guards to that extent that he was sent out one afternoon to assist one of the guards in overseeing the labor of a squad working in a stone quarry near by. Taking advantage of a time when the guard was a few hundred feet away, the other convicts knocked Johnson down and tried to get away. He got up, however, and interested them till the guard got to him and the escape was prevented. Johnson waited till all was secure again, and then fainted from loss of blood occasioned by a scalp wound over which he had a long fight afterward with erysipelas.

This was all lucky for me, and when I presented the petition to the Governor I had a strong case, made more so by the heroic action of a man who had been unjustly condemned.

There is but little more to tell. The Governor intimated that he would take favorable action upon the petition, but he wanted time. My great anxiety, as I told him, was to get the pardon in time so that Johnson could spend his

Christmas in freedom. I had seen him frequently, and he was pale and thin to emaciation. He could not live long if he remained where he was. I spoke earnestly of his good character since his incarceration, and the Governor promised prompt action. But he was called away in December and I feared that he might, in the rush and pressure of other business, forget the case of Johnson till after the holidays. So I telegraphed him and made his life a burden to him till the afternoon of the 24th, when the 4:50 train brought the pardon. In my poor, weak way I have been in the habit for some years of making Christmas presents, but nothing that could be bought with money ever made me a happier donor or donee than the simple act of giving to William Johnson four years of freedom which he did not look for.

I went away to spend my own Christmas, but not till I had given Johnson a few dollars' to help him get another start, and had made him promise to write me how he got along. And so that to me was a memorable and a joyous Christmas, for I had made myself happy by making others happy.   BILL NYE.

P. S.—Perhaps I ought not to close this account so abruptly as I have done, for the reader will naturally ask whether Johnson ever wrote me, as he said he would. I only received one letter from him, and that I found when I got back, a few days after Christmas. It was quite characteristic, and read as follows :

                    " Laramy the twenty-fitt dec.

FRENT NIE.

" When you get this Letter i will Be in A nuther tearritory whare the weekid seize from trubbling & the weery air at Reast excoose my Poor writing i refer above to the tearritory of Utaw where i will begin Life A new & all will be fergott.

" I hop god wil Reward you In Caise i Shood not Be Abel to Do so.

" You have Bin a good frent off me and so I am shure you will enjoy to heer of my success i hope the slooth hounds of Justiss will not try to folly me for it will be worse than Useles as i hav a damsite better team than i had Before.

" It is the Sheariff's team wich i have got & his name is denis, tel the Governer to Parden me if i have seeamed Rude i shall go to some new Plais whare i will not be Looked upon with Suchpishion wishing you a mary Crissmus hapy new year and April Fool i will Close from your tru Frent

"BIL JOHNSON."

From Delphi to Camden.

I.

From Delphi to Camden—little Hoosier towns,—
But here were classic meadows, blooming dales and downs
And here were grassy pastures, dewy as the leas
Trampled over by the trains of royal pageantries.

And here the winding highway loitered through the shade
Of the hazel-covert, where, in ambuscade,
Loomed the larch and linden, and the green-wood tree
Under which bold Robin Hood loud hallooed to me!

Here the stir and riot of the busy day,
Dwindled to the quiet of the breath of May;
Gurgling brooks, and ridges lily-marged, and spanned
By the rustic bridges found in Wonderland!

II.

From Delphi to Camden—from Camden back again!—
And now the night was on us, and the lightning and the rain;
And still the way was wondrous with the flash of hill and
              plain,—
The stars like printed asterisks—the moon a murky stain!
And I thought of tragic idyl, and of flight and hot pursuit,
And the jingle of the bridle, and cuirass, and spur on boot,
As our horses' hooves struck showers from the flinty
              bowlders set
In freshet ways with writhing reed and drowning violet.

And we passed beleaguered castles, with their battlements
              a-frown;
Where a tree fell in the forest was a turret toppled down;
While my master and commander—the brave knight I
              galloped with
On this reckless road to ruin, or to fame, was—Dr. Smith!

The Grammatical Boy

S O M E T I M E S a sad, homesick feeling comes over me, when I compare the prevailing style of anecdote and school literature with the old McGuffey brand, so well known thirty years ago. Today our juvenile literature, it seems to me, is so transparent, so easy to understand that I am not surprised to learn that the rising generation shows signs of lawlessness.

Boys today do not use the respectful language and large, luxuriant words that they did when Mr. McGuffey used to stand around and report their conversations for his justly celebrated school reader. It is disagreeable to think of, but it is none the less true, and for one I think we should face the facts.

I ask the careful student of school literature to compare the following selection, which I have written myself with great care, and arranged with special reference to the matter of choice and difficult words, with the flippant and commonplace terms used in the average school book of today.

One day as George Pillgarlic was going to his tasks, and while passing through the wood, he spied a tall man approaching in an opposite direction along the highway.

"Ah!" thought George, in a low, mellow tone of voice, "whom have we here?"

"Good morning, my fine fellow," exclaimed the stranger, pleasantly. "Do you reside in this locality?"

"Indeed I do," retorted George, cheerily, doffing his cap. "In yonder cottage, near the glen, my widowed mother and her thirteen children dwell with me."

"And is your father dead?" exclaimed the man, with a rising inflection.

"Extremely so," murmured the lad, "and, oh, sir, that is why my poor mother is a widow."

"And how did your papa die?" asked the man, as he thoughtfully stood on the other foot awhile.

"Alas! sir," said George, as a large hot tear stole down his pale cheek, and fell with a loud report on the warty surface of his bare foot, "he was lost at sea in a bitter gale. The good ship foundered two years ago last Christmastide, and father was foundered at the same time. No one knew of the loss of the ship and that the crew was drowned until the next spring, and it was then too late."

"And what is your age, my fine fellow?" quoth the stranger.

"If I live till next October," said the boy, in a declamatory tone of voice suitable for a Second Reader, "I will be seven years of age."

"And who provides for your mother and her large family of children?" queried the man.

"Indeed I do, sir," replied George, in a shrill tone. "I toil, oh, so hard, sir, for we are very, very poor, and since my elder sister, Ann, was married and brought her husband home to live with us, I have to toil more assiduously than heretofore."

"And by what means do you obtain a livelihood?" exclaimed the man, in slowly measured and grammatical words.

"By digging wells, kind sir," replied George, picking up a tired ant as he spoke and stroking it on the back. "I have a good education, and so I am able to dig wells as well as a man. I do this day-times and take in washing at night. In this way I am enabled barely to maintain our family in a precarious manner; but, oh, sir, should my other sisters marry, I fear that some of my brothers-in-law would have to suffer."

"And do you not fear the deadly fire-damp?" asked the stranger in an earnest tone.

"Not by a damp sight," answered George, with a low gurgling laugh, for he was a great wag.

"You are indeed a brave lad," exclaimed the stranger, as he repressed a smile. "And do you not at times become very weary and wish for other ways of passing your time?"

"Indeed I do, sir," said the lad. "I would fain run and romp and be gay like other boys, but I must engage in constant manual exercise, or we will have no bread to eat, and I have not seen a pie since papa perished in the moist and moaning sea."

"And what if I were to tell you that your papa did not perish at sea, but was saved from a humid grave?" asked the stranger in pleasing tones.

"Ah, sir," exclaimed George, in a genteel manner, again doffing his cap, "I am too polite to tell you what I would say, and beside, sir, you are much larger than I am."

"But, my brave lad," said the man in low musical tones, "do you not know me, Georgie? Oh, George!"

"I must say," replied George, "that you have the advantage of me. Whilst I may have met you before, I cannot at this moment place you, sir."

"My son! oh my son!" murmured the man, at the same time taking a large strawberry mark out of his valise and showing it to the lad. "Do you not recognize your parent on your father's side? When our good ship went to the bottom, all perished save me. I swam several miles through the billows, and at last, utterly exhausted, gave up all hope of life. Suddenly I stepped on something hard. It was the United States.

"And now, my brave boy," exclaimed the man with great glee, "see what I have brought for you." It was but the work of a moment to unclasp from a shawl-strap which he held in his hand and present to George's astonished gaze a large 40-cent water-melon, which until now had been concealed by the shawl-strap.

HIS CRAZY BONE.

His Crazy-Bone.

The man that struck his crazy-bone,
  All suddenly jerked up one foot
  And hopped three vivid hops, and put
    His elbow straight before him—then
Flashed white as pallid Parian stone,
  And clinched his eyes, and hopped again.

He spake no word—he made no moan—
  He muttered no invective—but
  Just gripped his eyelids tighter shut,
    And as the the world whizzed past him then,
He only knew his crazy-bone
  Was stricken—so—he hopped again.

The Chemist of the Carolinas.

ASHEVILLE, N. C. Dec. 13.—Last week I went out into the mountains for the purpose of securing a holly tree with red berries on it for Yuletide. I had noticed in all my pictures of Christmas festivities in England that the holly, with cranberries on it, constituted the background of Yuletide. A Yuletide in England without a holly bough and a little mistletoe in it wouldn't be worth half price. Here these vegetables grow in great profusion, owing to the equable climate, and so the holly tree is within the reach of all.

I resolved to secure one personally, so I sped away into the mountains where, in less than the time it takes to tell it, I had succeeded in finding a holly tree and losing myself. It is a very solemn sensation to feel that you are lost, and that before you can be found something is liable to happen to the universe.

I wandered aimlessly about for half an hour, hoping that I would be missed in society and some one sent in search of me. I was just about to give up in despair and sink down on a bed of moss with the idea of shuffling off six or seven feet of mortal coil when, a few rods away, I saw a blue smoke issuing from the side of the mountain and rising toward the sky. I went rapidly towards it and found it to be a plain dugout with a dirt floor. I entered and cast myself upon a rude nail keg, allowing my feet to remain suspended at the lower end of my legs, an attitude which I frequently affect when fatigued.

The place was not occupied at the time I entered, though there was a fire and things looked as though the owner had not been long absent. It seemed to be a kind of laboratory, for I could see here and there the earmarks of the chemist. I feared at first that it was a bomb fac-

MERRY CHRISTMAS K.D—J JADNE.

tory, but as I could not see any of these implements in a perfected state I decided that it was safe and waited for the owner to arrive.

After a time I heard a low guttural footstep approaching up the hill. I went to the door and exclaimed to the proprietor as he came, " Merry Christmas, Colonel."

" Merry Christmas be d——d!" said he in the same bantering tone. " What in three dashes, two hyphens and an asthonisher do you want here, you double-dashed and double-blanketed blank to dash and return !!''

The wording here is my own, but it gives an idea of the way the conversation was drifting. You can see by his manner that literary people are not alone in being surly, irritable and unreasonable.

So I humored him and spoke kindly to him and smoothed down his ruffled plumage with my gay badinage, for he wore a shawl and you can never tell whether a man wearing a shawl is armed or not. I give herewith a view of this chemist as he appeared on the morning I met him.

. It will be noticed that he was a man about medium height with clear-cut features and hair and retreating brisket. His hair was dark and hung in great waves which seemed to have caught the sunlight and retained it together with a great many other atmospheric phenomena. He wore a straw hat, such as I once saw Horace Greeley catch grasshoppers in, on the banks of the Kinnickinnick, just before he caught a small trout.

I spent some time with him watching him as he made his various experiments. Finally, he showed me a new beverage that he had been engaged in perfecting. It was inclosed in a dark brown stone receptacle and was held in place by a common corn-cob stopper. I took some of it in order to show that I confided in him. I do not remember anything else distinctly. The fumes of this drink went at once to my brain, where it had what might be termed a complete walkover.

I now have no hesitation in saying that the fluid must have been alcoholic in its nature, for when I regained my consciousness I was extremely elsewhere. I found myself on a road which seemed to lead in two opposite directions and my mind was very much confused.

I hardly know how I got home, but I finally did get there, accompanied by a strong leaning towards Prohibition. A few days ago I received the following letter:

Sir: I at first thought when I saw you at my laboratory the other day that you was a low, inquisitive cuss and so I spoke to you in harsh tones and reproached you and upbraided you by calling you everything I could lay my tongue to, but since then I have concluded that you didn't know any better.

You said to me that you found my place by seeing the smoke coming out of the chimbley; that has given me an idea that you might know something about what's called a smoke consumer of which I have heard. I am doing a fair business, but I am a good deal pestered, as you might say, by people who come in on me when I do not want to mingle in society. A man in the chemist business cannot succeed if he is all the time interrupted by Tom, Dick and Harry coming in on him when he is in the middle of an experiment.

I am engaged in making a remedy for which there is a great demand, but its manufacture is regarded with suspicion by United States officials who want to be considered zealous. Rather than be drawn into any difficulty with these people, I have always courted retirement and avoided the busy haunts of men. Still some strolling idiot or other will occasionally see the smoke from my little home and drop in on me.

Could you find out about this smoke consumer and see what the price would be and let me know as soon as possible?

If you could do so I can be of great service to you. Leave the letter under the big stone where you found yourself the other day when you came out of your trance. I call it a trance because this letter might fall into the hands of your family. If you will find out about this smoke consumer and leave the information where I have told you you will find on the following day a large jug of mountain dew in the same place that will make your hair grow and give a roseate hue to your otherwise gloomy life.

Do not try to come here again. It might compromise me. A man in your position may not have anything to risk, but with me it is different. My unsullied reputation is all I have to bequeath to my children. If you come often there will not be enough of it left to go around, as I have a large family.

If you hear of anybody that wants to trade a good double-barrel shotgun for a small portable worm and retort that is too small for my business, I can give him a good trade on it if he will let you know. This is a good machine for experimental purposes, and being no larger than a Babcock fire-extinguisher it can be readily conveyed to a place of safety at a very rapid rate.

You might say to your friends that we shall try in the future as we have in the past to keep up the standard of our goods, so as to merit a continued patronage.

Citizens of the United States, or those who have declared their intention to become such, will always be welcome at our works, provided they are not office-holders in any capacity.   We have no use for those who are in any way connected with the public teat.   I. B. MOONSHINE.

Dictated letter.

I hope that any one will feel perfectly free to address me in relation to anything referred to in the above letter. All communications containing remittances will be regarded as strictly confidential.

Craqueodoom

The Crankadox leaned o'er the edge of the moon
    And wistfully gazed on the sea
Where the Gryxabodill madly whistled a tune
    To the air of " Ti-fol-de-ding-dee."
The quavering shriek of the Fliupthecreek
    Was fitfully wafted afar
To the Queen of the Wunks as she powdered her cheek
    With the pulverized rays of a star.

The Gool closed his ear on the voice of the Grig,
    And his heart it grew heavy as lead
As he marked the Baldekin adjusting his wig
    On the opposite side of his head;
And the air it grew chill as the Gryxabodill
    Raised his dank, dripping fins to the skies,
To plead with the Plunk for the use of her bill
    To pick the tears out of his eyes.

The ghost of the Zhack flitted by in a trance;
    And the Squidjum hid under a tub
As he heard the loud hooves of the Hooken advance
    With a rub-a-dub-dub-a-dub dub!
And the Crankadox cried as he laid down and died,
    " My fate there is none to bewail!"
While the Queen of the Wunks drifted over the tide
    With a long piece of crape to her tail.

Prying Open the Future.

"Ring the bell and the door will open," is the remark made by a small label over a bell-handle in Third avenue, near Eighteenth street, where Mme. La Foy reads the past, present and future at so much per read. Love, marriage, divorce, illness, speculation and sickness are there handled with the utmost impunity by "Mme. La Foy, the famous

scientific astrologist," who has monkeyed with the planets for twenty years, and if she wanted any information has "read it in the stars."

I rang the bell the other day to see if the door would open. It did so after considerable delay, and a pimply boy in knee pants showed me upstairs into the waiting-room. After a while I was removed to the consultation-room, where

Mme. La Foy, seated behind a small oil-cloth covered table, rakes up old personalities and pries into the future at cut rates.

Skirmishing about among the planets for twenty years involves a great deal of fatigue and exposure, to say nothing of the night work, and so Mme. La Foy has the air of one who has put in a very busy life. She is as familiar with planets though as you or I might be with our own family, and calls them by their first names. She would know Jupiter, Venus, Saturn, Adonis or any of the other fixed stars the darkest night that ever blew.

"Mme. La Foy De Graw," said I, bowing with the easy grace of a gentleman of the old school, "would you mind peering into the future for me about a half dollar's worth, not necessarily for publication, et cetera."

"Certainly not. What would you like to know?"

"Why, I want to know all I can for the money," I said in a bantering tone. "Of course I do not wish to know what I already know. It is what I do not now know that I desire to know. Tell me what I do not know, Madame. I will detain you but a moment."

She gave me back my large, round half dollar and told me that she was already weary. She asked me to excuse her. She was willing to unveil the future to me in her poor, weak way, but she could not guarantee to let a large flood of light into the darkened basement of a benighted mind for half a dollar.

"You can tell me what year and on what day of what month you were born," said Mme. La Foy, "and I will outline your life to you. I generally require a lock of the hair, but in your case we will dispense with it."

I told her when I was born and the circumstances as well as I could recall them.

"This brings you under Venus, Mercury and Mars. These three planets were in conjunction at the time of your birth. You were born when the sign was wrong and you have had more or less trouble ever since. Had you been born when the sign was in the head or the heart, instead of the feet, you would not have spread out over the ground so much.

"Your health is very good, as is the health of those generally who are born under the same auspices that you were. People who are born under the reign of the crab are apt to be cancerous. You, however, have great lung power and wonderful gastric possibilities. Yet, at times, you would be easily upset. A strong cyclone that would unroof a courthouse or tip over a through train would also upset you, in spite of your broad, firm feet if the wind got behind one of your ears.

"You will be married early and you will be very happy, though your wife will not enjoy herself very much. Your wife will be much happier during her second marriage.

"You will prosper better in business matters without forming any partnerships. Do not go into partnership with a small, dark man who has neuralgia and a fine yacht. He has abundant means, but he will go through you like an electric shock.

"Tuesdays and Saturdays will be your most fortunate days on which to borrow money of men with light hair. Mondays and Thursdays will be your best days for approaching dark men.

"Look out for a low-sot man accompanied by an office cat, both of whom are engaged in the newspaper business. He is crafty and bald-headed on his father's side. He prints the only paper that contains the full text of his speeches at testimonials and dinners given to other people. Do not loan him money on any account.

"You would succeed well as a musician or an inventor, but you would not do well as a poet. You have all the keen sensibility and strong passion of a poet, but you haven't the hair. Do not try poesy.

"In. the future I see you very prosperous. You are on the lecture platform speaking. Large crowds of people are jostling each other at the box-office and trying to get their money back.

"Then I see you riding behind a flexible horse that must have cost a large sum of money. You are smoking a cigar that has never been in use before. Then Venus bisects the orbit of Mars and I see you going home with your head tied up in the lap robe, you and your spirited horse in the same ambulance."

"But do you see anything for me in the future, Mme. La Foy?" I asked, taking my feet off the table, the better to watch her features; "anything that would seem to indicate political preferment, a reward for past services to my country, as it were?"

"No, not clearly. But wait a moment. Your horoscope begins to get a little more intelligent. I see you at the door of the Senate Chamber. You are counting over your money and looking sadly at a schedule of prices. Then you turn sorrowfully away and decide to buy a seat in the House instead. Many years after I see you in the Senate. You are there day after day attending to your duties. You are there early, before any one else, and I see you pacing back and forth, up and down the aisles, sweeping out the Senate Chamber and dusting off the seats and rejuvenating the cuspidors."

"Does this horoscope which you are using this season give you any idea as to whether money matters will be scarce with me next week or otherwise, and if so what I had better do about it?"

"Towards the last of the week you will experience considerable monetary prostration, but just as you have become despondent, at the very tail end of the week, the horizon will clear up and a slight, dark gentleman, with wide trousers, who is a total stranger to you, will loan you quite a sum of money, with the understanding that it is to be repaid on Monday."

"Then you would not advise me to go to Coney Island until the week after next?"

"Certainly not."

"Would it be etiquette in dancing a quadrille to swing a young person of the opposite sex twice round at a select party when you are but slightly acquainted, but feel quite confident that her partner is unarmed?"

"Yes."

"Does your horoscope tell a person what to do with raspberry jelly that will not jell?"

"No, not at the present prices."

"So you predict an early marriage, with threatening weather and strong prevailing easterly winds along the Gulf States?"

"Yes, sir."

"And is there no way that this early marriage may be evaded?"

"No, not unless you put it off till later in life."

"Thank you," I said, rising and looking out the window over a broad sweep of undulating alley and wind-swept roofing, "and now, how much are you out on this?"

"Sir!"

"What's the damage?"

"Oh, one dollar."

"But don't you advertise to read the past, present and future for fifty cents?"

"Well, that is where a person has had other information before in his life and has some knowledge to begin with; but where I fill up a vacant mind entirely and store it with facts of all kinds and stock it up so that it can do business for itself, I charge a dollar. I cannot thoroughly refit and refurnish a mental tenement from the ground up for fifty cents."

I do not think we have as good "Astrologists" now as we used to have. Astrologists cannot crawl under the tent and pry into the future as they could three or four thousand years ago.

Mr. Silberberg.

I like me yet dot leedle chile
   Vich climb my lap up in to-
     day,
   Unt took my cheap cigair
     avay,
Unt laugh and kiss me purty-
    whvile,—
   Possescially I like dose mout'
   Vich taste his moder's like
     —unt so,
  Off my cigair it gone glean out
    —Yust let it go!

Vat I caire den for anyding?
   Der paper schlip out fon my
    hand,

And all my odvairtizement stand,
Mitout new changements boddering;
   I only dink—I have me dis
     Von leedle boy to pet unt love
   Unt play me vit, unt hug unt kiss—
     Unt dot's enough!

Der plans unt pairposes I vear
   Out in der vorld all fades avay;
   Unt vit der beeznid of der day
I got me den no time to spare;
   Der caires of trade vas caires no more—
     Dem cash accounds dey dodge me by,
   Unt vit my chile I roll der floor,
     Unt laugh unt gry!

Ah! frient! dem childens is der ones
   Dot got some happy times—you bet!—
   Dot's vy ven I been growed up yet
I vish I vould been leedle vonce!
   Unt ven dot leetle roozter tries
     Dem baby-tricks I used to do,
   My mout it vater, unt my eyes
     Dey vater too!

Unt all der summertime unt spring
   Of childhood it come back to me,
   So dot it vas a dream I see
Ven I yust look at anyding,
   Unt ven dot leedle boy run by,
     I dink "dot's me," fon hour to hour
   Schtill chasing yet dose butterfly
     Fon flower to flower!

Oxpose I vas lots money vairt,
   Mit blenty schtone-front schtore to rent,
   Unt mor'gages at twelf per-cent,

Unt diamonds in my ruffled shairt,—
  I make a'signment of all dot,
    Unt tairn it over mit a schmile,
   Obber you please—but don'd forgot,
    I keep dot chile!

Spirits at Home.

There was Father, and Mother, and Emmy, and Jane,
  And Lou, and Ellen, and John and me —
And father was killed in the war, and Lou
She died of consumption, and John did too,
  And Emmy she went with the pleurisy.

Father believed in 'em all his life —
  But Mother, at first, she'd shake her head —
Till after the battle of Champion Hill,
When many a flag in the winder-sill
  Had crape mixed in with the white and red!

I used to doubt 'em myself till then —
  But me and Mother was satisfied
When Ellen she set, and Father came
And rapped " God bless you ! " and Mother's name,
  And " The flag's up here ! "   And we just all cried!

Used to come often after that,
  And talk to us—just as he used to do,
Pleasantest kind !   And once, for John,
He said he was " lonesome but wouldn't let on —
  Fear mother would worry, and Emmy and Lou."

But Lou was the bravest girl on earth —
  For all she never was hale and strong
She'd have her fun !   With her voice clean lost
She'd laugh and joke us that when she crossed
  To father, *we'd* all come taggin' along!

Died — just that way!   And the raps was thick
   That night, as they often since occur,
Extry loud.   And when Lou got back
She said it was Father and her — and " whack! "
   She tuck the table — and we knowed *her !*

John and Emmy, in five years more,
   Both had went. — And it seemed like fate ! —
For the old home it burnt down, — but Jane
And me and Ellen we built again
   The new house, here, on the old estate.

·And a happier family I don't know
   Of anywheres — unless its *them* —
Father, with all his love for Lou,
And her there with him, and healthy, too,
   And laughin', with John and little Em.

And, first we moved in the new house here,
   They all dropped in for a long pow-wow.
" We like your buildin', of course," Lou said, —
" But wouldn't swop with you to save your head —
   For *we* live in the ghost of the old house, now! "

Healthy but out of the Race

In an interview which I have just had with myself, I have positively stated, and now repeat, that at neither the St. Louis nor Chicago Convention will my name be presented as a candidate.

But my health is bully.

We are upon the threshold of a most bitter and acrimonious fight. Great wisdom and foresight are needed at this hour, and the true patriot will forget himself and his own interests in his great yearning for the good of his common country and the success of his party. What we need at this time is a leader whose name will not be presented at the convention but whose health is good.

No one has a fuller or better conception of the great duties of the hour than I. How clearly to my mind are the duties of the American citizen outlined today! I have never seen with clearer, keener vision the great needs of

my country, and my pores have never been more open.
Four years ago I was in some doubt relative to certain
important questions which now are clearly and satisfactorily
settled in my mind.   I hesitated then where now I am fully
established, and my tongue was coated in the morning when
I arose, whereas now I bound lightly from bed, kick out a
window, climb to the roof by means of the fire-escape and
there rehearse speeches which I will make this fall in case
it should be discovered at either of the conventions that my
name alone can heal the rupture in the party and prevent
its works from falling out.

I think my voice is better also than it was either four,
eight, twelve or sixteen years ago, and it does not tire me so
much to think of things to say from the tail-gate of a train
as it did when I first began to refrain from presenting my
name to conventions.

According to my notion, our candidate should be a plain
man, a magnetic but hairless patriot, who should be suddenly thought of by a majority of the convention and
nominated by acclamation.   He should not be a hide-bound
politician, but on the contrary he should be greatly startled,
while down cellar sprouting potatoes, to learn that he has
been nominated.   That's the kind of man who always sur-
prises everybody with his sagacity when an emergency
arises.

In going down my cellar stairs the committee will do
well to avoid stepping on a large and venomous dog who
sleeps on the top stair.   Or I will tie him in the barn if I
can be informed when I am liable to be startled.

I have always thought that the neatest method of calling
a man to public life was the one adopted some years since
in the case of Cincinnatus.   He was one day breaking a
pair of nervous red steers in the north field.   It was a hot
day in July, and he was trying to summer fallow a piece
of ground where the jimson weeds grew seven feet high.

The plough would not scour, and the steers had turned the yoke twice on him. Cincinnatus had hung his toga on a tamarac pole to strike a furrow by, and hadn't succeeded in getting the plough in more than twice in going across. Dressing as he did in the Roman costume of 458 B. C., the blackberry vines had scratched his massive legs till they were a sight to behold. He had scourged Old Bright and twisted the tail of Bolly till he was sick at heart. All through the long afternoon, wearing a hot, rusty helmet with rabbit-skin ear tabs he had toiled on, when suddenly a majority of the Roman voters climbed over the fence and asked him to become dictator in place of Spurius Melius.

Cincinnatus.

Putting on his toga and buckling an old hame strap around his loins he said: "Gentlemen, if you will wait till I go to the house and get some vaseline on my limbs I will do your dictating for you as low as you have ever had it done." He then left his team standing in the furrow while he served his country in an official capacity for a little over twenty-nine years, after which he went back and resumed his farming.

Though 2,300 years have since passed away and histor-
ians have been busy with that epoch ever since, no one has
yet discovered the methods by which Cincinnatus organized
and executed this, the most successful " People's Movement " of which we are informed.

The great trouble with the modern boom is that it is too
precocious. It knows more before it gets its clothes on
than the nurse, the physician and its parents. It then
dies before the sap starts in the maple forests.

My object in writing this letter is largely to tone down and keep in check any popular movement in my behalf until the weather is more settled. A season-cracked boom is a thing I despise.

I inclose my picture, however, which shows that I am so healthy that it keeps me awake nights. I go about the house singing all the time and playing pranks on my grandparents. My eye dances with ill-concealed merriment, and my conversation is just as sparkling as it can be.

I believe that during this campaign we should lay aside politics so far as possible and unite on an unknown, homely, but sparkling man. Let us lay aside all race prejudices and old party feeling and elect a magnetic chump who does not look so very well, but who feels first rate.

Towards the middle of June I shall go away to an obscure place where I cannot be reached. My mail will be forwarded to me by a gentleman who knows how I feel in relation to the wants and needs of the country.

To those who have prospered during the past twenty years let me say they owe it to the perpetuation of the principles and institutions towards the establishment and maintenance of which I have given the best energies of my life. To those who have been unfortunate let me say frankly that they owe it to themselves.

I have never had less malaria or despondency in my system than I have this spring. My cheeks have a delicate bloom on them like a russet apple, and my step is light and elastic. In the morning I arise from my couch and, touching a concealed spring, it becomes an upright piano. I then bathe in a low divan which contains a jointed tank. I then sing until interfered with by property owners and tax-payers who reside near by. After a light breakfast of calf's liver and custard pie I go into the reception-room and wait for people to come and feel my pulse. In the afternoon I lie down on a lounge for two or three hours, wondering in what way I

can endear myself to the laboring man.   I then dine heartily
at my club.   In the evening I go to see the amateurs play
" Pygmalion and Galatea."   As I remain till the play is
over, any one can see that I am a very robust man.   After
I get home I write two or three thousand words in my diary.
I then insert myself into the bosom of my piano and sleep,
having first removed my clothes and ironed my trousers for
future reference.

In closing, let me urge one and all to renewed effort.
The prospects for a speedy and unqualified victory at the
polls were never more roseate.   Let us select a man upon
whom we can all unite, a man who has no venom in him,
a man who has successfully defied and trampled on the
infamous Interstate Commerce act, a man who, though in
the full flush and pride and bloom and fluff of life's meridian, still disdains to present his name to the convention.

Lines.

Portentous sound! mysteriously
vast
And awful in the grandeur of
refrain
That lifts the listener's hair, as
it swells past,
And pours in turbid currents
down the lane.

The small boy at the woodpile,
in a dream,
Slow trails the meat-
rind o'er the listless
saw;

The chickens roosting o'er him on the beam
Uplift their drowsy heads, with cootered awe.

The "Gung-oigh" of the pump is strangely stilled;
  The smoke-house door bangs once emphatic'ly,
Then bangs no more, but leaves the silence filled
  With one lorn plaint's despotic minstrelsy.

Yet I would join thy sorrowing madrigal,
  Most melancholy cow, and sing of thee
Full-hearted through my tears, for, after all,
  Tis very kine of you to sing for me.

Me and Mary.

All my feelin's, in the spring,
   Gits so blame contrary
I can't think of anything
   Only me and Mary!
" Me and Mary! " all the time,
   " Me and Mary! " like a rhyme
Keeps a-dingin' on till I'm
   Sick o' "Me and Mary!"

" Me and Mary! · Ef us two
   Only was together—
Playin' like we used to do
   In the Aprile weather !"
All the night and all the day
   I keep wishin' thataway
Till I'm gittin' old and gray
   Jist on " Me and Mary!"

Muddy yit along the pike
   Sense the winter's freezin'
And the orchard's backard-like
   Bloomin' out this season ;
Only heerd one bluebird yit—
Nary robin er tomtit ;
What's the how and why of it?
   S'pect its " Me and Mary !"

Me and Mary liked the birds—
    That is, Mary sorto'
Liked them first, and afterwerds
    W'y I thought I orto.
And them birds—ef Mary stood
Right here with me as she should—
They'd be singin', them birds would,
    All fer me and Mary!

Birds er not, I'm hop'in' some
    I can git to plowin':
Ef the sun'll only come,
    And the Lord allowin',
Guess to-morry I'll turn in
And git down to work agin:
This here loaferin' won't win;
    Not fer me and Mary!

Fer a man that loves, like me,
    And's afeard to name it,
Till some other feller, he
    Gits the girl—dad-shame-it!
Wet er dry, er clouds er sun—
Winter gone, er jist begun—
Out-door work fer me er none,
    No more " Me and Mary!"

Niagara Falls From the Nye Side.

I visited Walton, N. Y., last week, a beautiful town in the flank of the Catskills, at the head of the Delaware. It was there in that quiet and picturesque valley that the great philanthropist and ameliorator, Jay Gould, first attracted attention. He has a number of relatives there who note with pleasure the fact that Mr. Gould is not frittering away his means during his lifetime.

In the office of Mr. Nish, of Walton, there is a map of the county made by Jay Gould while in the surveying business, and several years before he became monarch of all he surveyed.

Mr. Gould also laid out the town of Walton. Since that he has laid out other towns, but in a different way. He also plotted other towns. Plotted to lay them out, I mean.

In Franklin there is an old wheelbarrow which Mr. Gould used on his early surveying trips. In this he carried his surveying instruments, his night shirt and manicure set. Connected with the wheel there is an arrangement by which, at night, the young surveyor could tell at a glance, with the aid of a piece of red chalk and a barn door, just how far he had traveled during the day.

This instrument was no doubt the father of the pedometer and the cyclorama, just as the boy is frequently father to the man. It was also no doubt the *avant courier* of the Dutch clock now used on freight cabooses, which not only shows how far the car has traveled, but also the rate of speed for each mile, the average rainfall and whether the conductor has eaten onions during the day.

This instrument has worked quite a change in railroading since my time. Years ago I can remember when I used to ride in a caboose and enjoy myself, and before good fortune had made me the target of the alert and swift-flying whisk-broom of the palace car, it was my chief joy to catch a freight over the hill from Cheyenne, on the Mountain division. We were not due anywhere until the following day, and so at the top of the mountain we would cut off the caboose and let the train go on. We would then go into the glorious hills and gather sage-hens and cotton-tails. In the summer we would put in the afternoon catching trout in Dale Creek or gathering maiden-hair ferns in the bosky dells. Bosky dells were more plenty there at that time than they are now.

It was a delightful sensation to know that we could loll about in the glorious weather, secure a small string of stark, varnished trout with chapped backs, hanging aimlessly by one gill to a gory, willow stringer and then beat our train home by two hours by letting off the brakes and riding twenty miles in fifteen minutes.

But Mr. Gould saw that we were enjoying ourselves, and so he sat up nights to oppress us. The result is that the freight conductor has very little more fun now than Mr. Gould himself. All the enjoyment that the conductor of " Second Seven " has now is to pull up his train where it will keep the passengers of No. 5 going west from getting a view of the town. He can also, if he be on a night run, get under the window of a sleeping-car at about 1:35 A. M., and make a few desultory remarks about the delinquency of " Third Six " and the lassitude of Skinny Bates who is supposed to brake ahead on No. 11 going west. That is all the fun he has now.

I saw Niagara Falls on Thursday for the first time. The

sight is one long to be remembered. I did not go to the falls, but viewed them from the car window in all their might, majesty, power and dominion forever. N. B.—Dominion of Canada.

Niagara Falls plunges from a huge elevation by reason of its inability to remain on the sharp edge of a precipice several feet higher than the point to which the falls are now falling. This causes a noise to make its appearance, and a thick mist, composed of minute particles of wetness, rises to its full height and comes down again afterwards. Words are inadequate to show here, even with the aid of a large, powerful new press, the grandeur, what you may call the vertigo, of Niagara. Everybody from all over the world goes to see and listen to the remarks of this great fall. How convenient and pleasant it is to be a cataract like that and have people come in great crowds to see and hear you! How much better that is than to be a lecturer, for instance, and have to follow people to their homes in order to attract their attention!

Many people in the United States and Canada who were once as pure as the beautiful snow, have fallen, but they did not attract the attention that the fall of Niagara does.

For the benefit of those who may never have been able to witness Niagara Falls in winter, I give here a rough sketch of the magnificent spectacle as I saw it from the

American side.   From the Canadian side the aspect of the
falls is different, and the names on the cars are not the same,
but the effect on one of a sensitive nature is one of intense
awe.   I know that I cannot put so much of this awe into
a hurried sketch as I would like to.   In a crude drawing,
made while the train was in motion, and at a time when the
customs officer was showing the other passengers what I had

in my valise, of course I could not make a picture with
much sublimity in it, but I tried to make it as true to
nature as I could.

The officer said that I had nothing in my luggage that was liable to duty, but stated that I would need heavier underwear in Canada than the samples I had with me.

Toronto is a stirring city of 150,000 people, who are justly proud of her great prosperity. I only regretted that I could not stay there a long time.

I met a man in Cleveland, O., whose name was Macdonald. He was at the Weddell House, and talked freely with me about our country, asking me a great many questions about myself and where I lived and how I was prospering. While we were talking at one time he saw something in the paper which interested him and called him away. After he had gone I noticed the paragraph he had been reading, and saw that it spoke of a man named Macdonald who had recently arrived in town from New York, and who was introducing a new line of green goods.

I have often wondered what there is about my general appearance which seems to draw about me a cluster of green-goods men wherever I go. Is it the odor of new-mown hay, or the frank, open way in which I seem to measure the height of the loftiest buildings with my eye as I penetrate the busy haunts of men and throng the crowded marts of trade? Or do strangers suspect me of being a man of means?

In Cleveland I was rather indisposed, owing to the fact that I had been sitting up until 2 or 3 o'clock A. M. for several nights in order to miss early trains. I went to a physician, who said I was suffering from some new and attractive disease, which he could cope with in a day or two. I told him to cope. He prescribed a large 42-calibre capsule which he said contained medical properties. It might have contained theatrical properties and still had room left for a baby grand piano. I do not know why the capsule should be so popular. I would rather swallow a porcelain egg or a live turtle. Doctors claim that it is to prevent the bad taste of the medicines, but I have never yet participated in any medicine which was more disagreeable than the gluey shell of an adult capsule, which looks like an overgrown bott and tastes like a rancid nightmare.

I doubt the good taste of any one who will turn up his nose at castor-oil or quinine and yet meekly swallow a chrysalis with varnish on the outside.

Everywhere I go I find people who seem pleased with the manner in which I have succeeded in resembling the graphic pictures made to represent me in *The World*. I can truly say that I am not a vain man, but it is certainly pleasing and gratifying to be greeted by a glance of recognition and a yell of genuine delight from total strangers. Many have seemed to suppose that the massive and undraped head shown in these pictures was the result of artistic license or indolence and a general desire to evade the task of making hair. For such people the thrill of joy they feel when they discover that they have not been deceived is marked and genuine.

These pictures also stimulate the press of the country to try it themselves and to add other horrors which do not in any way interfere with the likeness, but at the same time encourages me to travel mostly by night.

"Curly Locks!"

"*Curly Locks! Curly Locks! wilt thou be mine?*
Thou shalt not wash the dishes, nor yet feed the swine,—
But sit on a cushion and sew a fine seam,
And feast upon strawberries, sugar and cream."

Curly Locks! Curly Locks! wilt thou be mine?
The throb of my heart is in every line,
And the pulse of a passion, as airy and glad
In its musical beat as the little Prince had!

Thou shalt not wash the dishes, nor yet feed the swine!—
O, I'll dapple thy hands with these kisses of mine
Till the pink of the nail of each finger shall be
As a little pet blush in full blossom for me.

But sit on a cushion and sew a fine seam,
And thou shalt have fabric as fair as a dream,—
The red of my veins, and the white of my love,
And the gold of my joy for the braiding thereof.

And feast upon strawberries, sugar and cream
From a service of silver, with jewels agleam, —
At thy feet will I bide, at thy beck will I rise,
And twinkle my soul in the night of thine eyes!

" *Curly Locks! Curly Locks! wilt thou be mine?*
Thou shalt not wash the dishes, nor yet feed the swine,
But sit on a cushion and sew a fine seam,
And feast upon strawberries, sugar and cream."

Lines on Turning Over a Pass.

OME newspaper men claim that they feel a great deal freer if they pay their fare.

That is true, no doubt; but too much freedom does not agree with me. It makes me lawless. I sometimes think that a little wholesome restriction is the best thing in the world for me. That is the reason I never murmur at the conditions on the back of an annual pass. Of course they restrict me from bringing suit against the road in case of death, but I don't mind that. In case of my death it is my intention to lay aside the cares and details of business and try to secure a change of scene and complete rest. People who think that after my demise I shall have nothing better to do than hang around the musty, tobacco-spattered corridors of a court-room and wait for a verdict of damages against a courteous railroad company do not thoroughly understand my true nature.

But the interstate-commerce bill does not shut out the employe! Acting upon this slight suggestion of hope, I wrote, a short time ago, to Mr. St. John, the genial and whole-souled general passenger agent of the Chicago, Rock Island & Pacific Railroad, as follows:

ASHEVILLE, N. C., Feb. 10, 1887.

E. St. John, G. P. A., C., R. I. & P. R'y, Chicago :

Dear Sir—Do you not desire an employe on your charming road? I do not know what it is to be an employe, for I was never in that condition, but I pant to be one now.

Of course I am ignorant of the duties of an employe, but I have always been a warm friend of your road and rejoiced in its success. How are your folks?   Yours truly,   COL. BILL NYE.

Day before yesterday I received the following note from General St. John, printed on a purple type-writer:

CHICAGO, Feb. 13, 1887.

Col. Bill Nye, Asheville, N. C.:

Sir—My folks are quite well. Yours truly,   E. ST. JOHN.

I also wrote to Gen. A. V. H. Carpenter, of the Milwaukee road, at the same time, for we had corresponded ome back and forth in the happy past. I wrote in about the following terms:

ASHEVILLE, N. C., Feb. 10, 1887.

A. V. H. Carpenter, G. P. A. C., M. & St. P. R'y, Milwaukee, Wis.:

Dear Sir—How are you fixed for employes this spring?

I feel like doing something of that kind and could give you some good endorsements from prominent people both at home and abroad.

What does an employe have to do?

If I can help your justly celebrated road any here in the South do not hesitate about mentioning it.

I am still quite lame in my left leg, which was broken in the cyclone, and cannot walk without great pain. Yours with kindest regards,

BILL NYE.

I have just received the following reply from Mr. Carpenter:

MILWAUKEE, Wis. Feb. 14, 1887.

Bill Nye, Esq., Asheville, N. C:

Dear Sir—You are too late. As I write this letter, there is a string of men extending from my office door clear down to the Soldiers' Home. All of them want to be employes. This crowd embraces the Senate and House of Representatives of the Wisconsin Legislature, State officials, judges, journalists, jurors, justices of the peace, orphans, overseers of highways, fish commissioners, pugilists, widows of pugilists,, unidentified orphans of pugilists, etc., etc., and they are all just about as well qualified to be employes as you are.

I suppose you would poultice a hot box with pounded ice, and so would they.

I am sorry to hear about your lame leg. The surgeon of our road says perhaps you do not use it enough.

Yours for the thorough enforcement of law,

A. V. H. CARPENTER. Per G.

Not having written to Mr. Hughitt of the Northwestern road for a long time, and fearing that he might think I had grown cold toward him, I wrote the following note on the 9th:

ASHEVILLE. N. C.. Feb. 9, 1887.

Marvin Hughitt, Second Vice-President and General Manager Chicago & Northwestern Railway, Chicago, Ill.

Dear Sir—Exuse me for not writing before. I did not wish to write you until I could do so in a bright and cheery manner, and for some weeks I have been the hot-bed of twenty-one Early Rose boils. It was extremely humorous without being funny. My enemies gloated over me in ghoulish glee.

I see by a recent statement in the press that your road has greatly increased in business. Do you feel the need of an employe? Any light employment that will be honorable without involving too much perspiration would be acceptable.

I am traveling about a good deal these days, and if I can do you any good as an agent or in referring to your smooth road-bed and the magnificent scenery along your line, I would be glad to regard that in the light of employment. Everywhere I go I hear your road very highly spoken of.

Yours truly, BILL NYE.

I shall write to some more roads in a few weeks. It seems to me there ought to be work for a man who is able and willing to be an employe.

That Night.

You and I, and that
    night, with its per-
    fume and glory!—
The scent of the lo-
    custs—the light of
    the moon,
And the violin weaving
    the waltzers a story,
Enmeshing their feet
    in the weft of the
    tune,
        Till their shadows uncertain,
        Reeled round on the curtain,
While under the trellis we drank in the
    June.

Soaked through with the midnight, the
    cedars were sleeping,

Their shadowy tresses outlined in the bright
Crystal, moon-smitten mists, where the fountain's heart
    leaping
  Forever, forever burst, full with delight;
    And its lisp on my spirit
    Fell faint as that near it
Whose love like a lily bloomed out in the night.

O your glove was an odorous sachet of blisses!
  The breath of your fan was a breeze from Cathay!
And the rose at your throat was a nest of spilled kisses!—
  And the music!—in fancy I hear it to-day,
    As I sit here, confessing
    Our secret, and blessing
My rival who found us, and waltzed you away.

The Truth about Methuselah.

E first meet Methuselah in the capacity of a son. At the age of sixty-five Enoch arose one night and telephoned his family physician to come over and assist him in meeting Methuselah. Day at last dawned on Enoch's happy home, and its first red rays lit up the still redder surface of the little stranger. For three hundred years Enoch and Methuselah jogged along together in the capacity of father and son. Then Enoch was suddenly cut down. It was at this time that little Methuselah first realized what it was to be an orphan. He could not at first realize that his father was dead. He could not understand why Enoch, with no inherited disease, should be shuffled off at the age of three hundred and sixty-five years. But the doctor said to Methuselah: "My son, you are indeed fatherless. I have done all I could, but it is useless. I have told Enoch many a time that if he went in swimming before the ice went out of the creek it would finally down him, but he thought he knew better than I did. He was a headstrong man, Enoch was. He sneered at me and alluded to me as a fresh young gosling, because he was three hundred years older than I was. He has received the reward of the willful, and verily the doom of the smart Aleck is his."

Methuselah now cast about him for some occupation which would take up his atttention and assuage his wild, passionate grief over the loss of his father. He entered into the walks of men and learned their ways. It was at this time that he learned the pernicious habit of using tobacco. We cannot wonder at it when we remember that he was now fatherless. He was at the mercy of the coarse, rough world. Possibly he learned the use of tobacco when he went away to attend business college after the death of his father. Be that as it may, the noxious weed certainly hastened his death, for six hundred years after this we find him a corpse!

Death is ever a surprise, even at the end of a long illness and after a ripe old age. To those who are near it seems abrupt; so to his grandchildren, some of whom survived him, his children having died of old age, the death of Methuselah came like a thunderbolt from a clear sky.

Methuselah succeeded in cording up more of a record, such as it was, than any other man of whom history informs us. Time, the tomb-builder and amateur mower came and leaned over the front yard and looked at Methuselah, and ran his thumb over the jagged edge of his scythe, and went away whistling a low refrain. He kept up this refrain business for nearly ten centuries, while

Methuselah continued to stand out amid the general wreck of men and nations.

Even as the young, strong mower going forth with his mower for to mow spareth the tall and drab hornet's nest and passeth by on the other side, so Time, with his Waterbury hour-glass and his overworked hay-knife over his shoulder, and his long Mormon whiskers, and his high sleek dome of thought with its gray lambrequin of hair around the base of it, mowed all around Methuselah and then passed on.

Methuselah decorated the graves of those who perished in a dozen different wars. He did not enlist himself, for over nine hundred years of his life he was exempt. He would go to the enlisting places and offer his services, and the officer would tell him to go home and encourage his grandchildren to go. Then Methuselah would sit around Noah's front steps, and smoke and criticise the conduct of the war, also the conduct of the enemy.

It is said of Methuselah that he never was the same man after his son Lamech died. He was greatly attached to Lamech, and, when he woke up one night to find his son purple in the face with membraneous croup, he could hardly realize that he might lose him. The idea of losing a boy who had just rounded the glorious morn of his 777th year had never occurred to him. But death loves a shining mark, and he garnered little Lammie and left Methuselah to mourn for a couple of centuries.

Methuselah finally got so that he couldn't sleep any later than 4 o'clock in the morning, and he didn't see how any one else could. The older he got, and the less valuable his time became, the earlier he would rise, so that he could get an early start. As the centuries filed slowly by, and Methuselah got to where all he had to do was to shuffle into his loose-fitting clothes and rest his gums on the top of a large slick-headed cane and mutter up the chimney, and then groan and extricate himself from his clothes again and retire, he rose earlier and earlier in the morning, and muttered more and more about the young folks sleeping away the best of the day, and he said he had no doubt that sleeping and snoring till breakfast time helped to carry off Lam. But one day old Father Time came along with a new scythe, and he drew the whetstone across it a few times, and rolled the sleeves of his red-flannel undergarment up over his warty elbows, and Mr. Methuselah passed on to that undiscovered country, with a ripe experience and a long, clean record.

We can almost fancy how the physicians, who had disagreed about his case all the way through, came and insisted on a post-mortem examination to prove which was right and what was really the matter with him. We can imagine how people went by shaking their heads and regretting that Methuselah should have tampered with tobacco when he knew that it affected his heart.

But he is gone. He lived to see his own promissory notes rise, flourish, acquire interest, pine away at last and finally outlaw. He acquired a large farm in the very heart of the county-seat, and refused to move or to plot, and called it Methuselah's addition. He came out in spring regularly for nine hundred years after he got too old to work out his poll-tax on the road, and put in his time telling the rising generation how to make a good road. Meantime other old people, who were almost one hundred years of age, moved away and went West where they would attract attention and command respect. There was actually no pleasure in getting old around where Methuselah was, and being ordered about and scolded and kept in the background by him.

So, when at last he died, people sighed and said: "Well, it was better for him to die before he got childish. It was best that he should die at a time when he knew it all. We can't help thinking what an acquisition Methuselah will be on the evergreen shore when he gets there, with all his ripe experience and his habits of early rising."

And the next morning after the funeral Methuselah's family did not get out of bed till nearly 9 o'clock.

A Black Hills Episode.

A little, warty, dried-up sort
O' lookin' chap 'at hadn't ort
A ben a-usin' round no bar,
With gents like us a-drinkin' thar!

And that idee occurred to me
The livin' minit 'at I see
The little cuss elbowin' in
To humor his besettin' sin.

There 're nothin' small in me at all,
But when I heer the rooster call
For shugar and a spoon, I says:
"Jest got in from the States, I guess."

He never 'peared as if he heerd,
But stood thar, wipin' uv his beard,
And smilin' to hisself as if
I'd been a-givin' him a stiff.

And I-says-I, a edgin' by
The bantam, and a-gazin' high
Above his plug—says I: "I knowed
A little feller onc't 'at blowed

"Around like you, and tuck his drinks
With shugar in—and *his* folks thinks
He's dead now—'cause we boxed and sent
The scraps back to the Settlement!"

  *   *   *   *   *

The boys tells me, 'at got to see

His *modus operandum*, he
Jest 'peared to come onjointed-like
Afore he ever struck a strike!

And I'll admit, the way he fit
Wuz dazzlin'—what I see uv hit;
And squarin' things up fair and fine,
Says I: "A little 'shug' in mine!"

The Rossville Lecture Course.

ROSSVILLE, Mich., March '87.—

OLKS up here at Rossville got up
  a lectur'-course ;
All the leadin' citizens they wus out
  in force ;
Met and talked at Williamses, and
  'greed to meet agin,
And helt another corkus when the
  next reports wuz in ;
Met agin at Samuelses ; and met
  agin at Moore's,
And Johnts he put the shutters up
  and jest barred the doors ! —
And yit, I'll jest be dagg-don'd ! ef didn't take a week
'Fore we'd settled where to write to git a man to speak !

Found out where the Bureau wus, and then and there
    agreed
To strike while the iron's hot, and foller up the lead.

Simp was secatary; so he tuck his pen in hand,
And ast what they'd tax us for the one on " Holy Land "—
" One of Colonel J. De-Koombs Abelust and Best
Lecturs," the circ'lar stated, " Give East er West! "
Wanted fifty dollars, and his kyar-fare to and from,
And Simp was hence instructed fer to write him not to
    come.

Then we talked and jawed around another week er so,
And writ the Bureau 'bout the town a-bein' sort o' slow
And fogey-like, and pore as dirt, and lackin' enterprise,
And ignornter'n any other 'cordin' to its size:
Till finally the Bureau said they'd send a cheaper man
Fer forty dollars, who would give "A Talk About Japan "—
" A regular Japanee hiss'f," the pamphlet claimed; and so,
Nobody knowed his languige, and of course we let him go!

Kindo' then let up a spell—but rallied onc't ag'in,
And writ to price a feller on what's called the " violin "—
A Swede, er Pole, er somepin—but no matter what he wus,
Doc Sifers said he'd heerd him, and he wusn't wuth a kuss!
And then we ast fer *Swingses* terms; and *Cook*, and
    Ingersoll—
And blame! ef forty dollars looked like anything at all!
And then *Burdette*, we tried fer him; and Bob he writ to
    say
He was busy writin' ortographts, and couldn't git away.

At last—along in Aprile—we signed to take this-here
Bill Nye of Californy, 'at was posted to appear
"The Humorestest Funny Man 'at Ever Jammed a Hall!"
So we made big preparations, and swep' out the church
    and all!

And night he wus to lectur', and the neighbors all was
   there,
And strangers packed along the aisles 'at come from ever'-
   where,
Committee got a telegrapht the preacher read, 'at run —
"Got off at Rossville, Indiany, 'stead of Michigun."

The Tar-heel Cow.

ASHEVILLE, N. C.,
December 9.—There is
no place in the United
States, so far as I know,
where the cow is more
versatile or ambidex-
trous, if I may be al-
lowed the use of a term
that is far above my
station in life, than here
in the mountains of
North Carolina, where
the obese 'possum and
the anonymous distiller
have their homes.

Not only is the Tar-
heel cow the author of
a pale, but athletic style
of butter, but in her
leisure hours she aids
in tilling the perpen-
dicular farm on the
hillside, or draws the
products to market. In
this way she contrives
to put in her time to
the best advantage, and
when she dies, it casts
a gloom over the com-
munity in which she
has resided.

The life of a North Carolina cow is indeed fraught with various changes and saturated with a zeal which is praiseworthy in the extreme. From the sunny days when she gambols through the beautiful valleys, inserting her black retrousse and perspiration-dotted nose into the blue grass from ear to ear, until at life's close, when every part and portion of her overworked system is turned into food, raiment or overcoat buttons, the life of a Tar-heel cow is one of intense activity.

Her girlhood is short, and almost before we have deemed her emancipated from calfhood herself we find her in the capacity of a mother. With the cares of maternity other demands are quickly made upon her. She is obliged to ostracize herself from society, and enter into the prosaic details of producing small, pallid globules of butter, the very pallor of which so thoroughly belies its lusty strength.

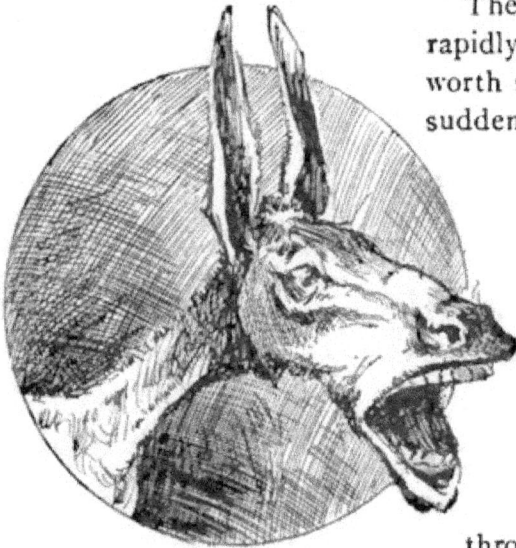

The butter she turns out rapidly until it begins to be worth something, when she suddenly suspends publication and begins to haul wood to market. In this great work she is assisted by the pearl-gray or ecru colored jackass of the tepid South. This animal has been referred to in the newspapers throughout the country, and yet he never ceases to be an object of the greatest interest.

Jackasses in the South are of two kinds, viz., male and female. Much as has been said of the jackass pro and con,

I do not remember ever to have seen the above statement in print before, and yet it is as trite as it is incontrovertible. In the Rocky mountains we call this animal the burro. There he packs bacon, flour and salt to the miners. The miners eat the bacon and flour, and with the salt they are enabled successfully to salt the mines.

The burro has a low, contralto voice which ought to have some machine oil on it. The voice of this animal is not unpleasant if he would pull some of the pathos out of it and make it more joyous.

Here the jackass at times becomes a co-worker with the cow in hauling tobacco and other necessaries of life into town, but he goes no further in the matter of assistance. He compels her to tread the cheese press alone and contributes nothing whatever in the way of assistance for the butter industry.

The North Carolina cow is frequently seen here driven double or single by means of a small rope line attached to a tall, emaciated gentleman, who is generally clothed with the divine right of suffrage, to which he adds a small pair of earbobbs during the holidays.

The cow is attached to each shaft and a small singletree, or swingletree, by means of a broad strap harness. She also wears a breeching, in which respect she frequently has the advantage of her escort.

I think I have never witnessed a sadder sight than that of a new milch cow, torn away from home and friends and kindred dear, descending a steep, mountain road at a rapid rate and striving in her poor, weak manner to keep out of the way of a small Jackson Democratic wagon loaded with a big hogshead full of tobacco. It seems to me so totally foreign to the nature of the cow to enter into the tobacco traffic, a line of business for which she can have no sympathy and in which she certainly can feel very little interest.

Tobacco of the very finest kind is produced here, and is

used mainly for smoking purposes. It is the highest-price tobacco produced in this country. A tobacco broker here yesterday showed me a large quantity of what he called export tobacco. It looks very much like other tobacco while growing.

He says that foreigners use a great deal of this kind. I am learning all about the tobacco industry while here, and as fast as I get hold of any new facts I will communicate them to the press. The newspapers of this country have done much for me, not only by publishing many pleasant things about me, but by refraining from publishing other things about me, and so I am glad to be able, now and then, to repay this kindness by furnishing information and facts for which I have no use myself, but which may be of incalculable value to the press.

As I write these lines I am informed that the snow is twenty-six inches deep here and four feet deep at High Point in this State. People who did not bring in their pomegranates last evening are bitterly bewailing their thoughtlessness today.

A great many people come here from various parts of the world, for the climate. When they have remained here for one winter, however, they decide to leave it where it is.

It is said that the climate here is very much like that of Turin. But I did not intend to go to Turin even before I heard about that.

Please send my paper to the same address, and if some one who knows a good remedy for chilblains will contribute it to these columns, I shall watch for it with great interest. Yours as here 2 4, BILL NYE.

P. S.—I should have said, relative to the cow of this State that if the owners would work their butter more and their cows less, they would confer a great boon on the consumer of both. B. N.

A Character.

I.

Swallowed up in gulfs of
  tho't—
Eye-glass   fixed—on—
  who knows what?
We but know he sees us
  not.

Chance upon him, here
  and there—
Base-ball park—Industri-
  al Fair—
Broadway—Long Branch
  —anywhere!

Even at the races,—yet
With his eye-glass tranced
  and set
On   some   dream-land
  minaret.

At the beach, the where,
  perchance—
Tenderest of eyes may glance
On the fitness of his pants.

Vain! all admiration—vain!
His mouth, o'er and o'er again,
Absently absorbs his cane.

Vain, as well, all tribute paid
To his morning coat, inlaid
With crossbars of every shade.

He is so oblivious, tho'
We played checkers to and fro
On his back—he would not know.

II.

So removed—illustrious—
Peace! kiss hands, and leave him thus,
He hath never need of us!

Come away! Enough! Let be!
Purest praise, to such as he,
Were as basest obloquy.

Vex no more that mind of his,
We, to him, are but as phizz
Unto pop that knows it is.

Haply, even as we prate
Of him HERE—in astral state—
Or jackastral—he, elate,

Brouses 'round, with sportive hops
In far fields of sphery crops,
Nibbling stars like clover-tops.

He, occult and psychic, may
Now be solving why to-day
Is not midnight.—But away!

Cease vain queries! Let us go!
Leave him all unfathomed.—Lo,
He can hear his whiskers grow.

The Diary of Darius T. Skinner.

"Fifth Avenue Hotel, New York, Dec. 31, 188 .—It hardly seems possible that I am here in New York, putting up at a hotel where it costs me $5 or $6 a day just simply to

exist. I came here from my far away-home entirely alone. I have no business here, but I simply desired to rub up against greatness for awhile.   I need polish, and I am smart enough to know it.

"I write this entry in my diary to explain who I am and to help identify myself in case I should come home to my room intoxicated some night and blow out the gas.

"The reason I am here is, that last summer while whacking bulls, which is really my business, I grub-staked Alonzo McReddy and forgot about it till I got back and the boys told me that Lon had struck a First National bank in the shape of the Sarah Waters claim.   He was then very low with mountain fever and so nobody felt like jumping the claim.   Saturday afternoon Alonzo passed away and left me the Sarah Waters.   That's the only sad thing about the whole business now.   I am raised from bull-whacking to affluence, but Alonzo is not here.   How we would take in the town together if he'd lived, for the Sarah Waters was enough to make us both well fixed.

"I can imagine Lon's look of surprise and pride as he looks over the outer battlements of the New Jerusalem and watches me paint the town.   Little did Lon think when I pulled out across the flat with my whiskers full of alkali dust and my cuticle full of raw agency whisky, that inside of a year I would be a nabob, wearing biled shirts every single day of my life, and clothes made specially for me.

"Life is full of sudden turns, and no one knows here in America where he'll be in two weeks from now.   I may be back there associating with greasers again as of yore and skinning the same bulls that I have heretofore skun.

"Last evening I went to see 'The Mikado,' a kind of singing theatre and Chinese walk-around.   It is what I would call no good.   It is acted out by different people who claim they are Chinamen, I reckon.   They teeter around on the stage and sing in the English language, but their clothes are peculiar.   A homely man, who played that he

was the lord high executioner and chairman of the vigilance
committee, wore a pair of wide, bandana pants, which came
off during the first act.   He was cool and collected, though,
and so caught them before it was everlasting too late.   He
held them on by one hand while he sang the rest of his piece,
and when he left the stage the audience heartlessly whooped
for him to come back.

"'The Mikado' is not funny or instructive as a general
thing, but last night it was accidently facetious.   It has too
much singing and not enough vocal music about it.   There
is also an overplus of conversation through the thing that
seems like talking at a mark for $2 a week.   It may be
owing to my simple ways, but 'The Mikado' is too rich for
my blood.

"We live well here at the Fifth Avenue.   The man that
owns the place puts two silver forks and a clean tablecloth
on my table every day, and the young fellows that pass the
grub around are so well dressed that it seems sassy and pre-
sumptious for me to bother them by asking them to bring
me stuff when I'd just as soon go and get it myself and
nothing else in the world to do.

"I told the waiter at my table yesterday that when he
got time I wished he would come up to my room and we
could have a game of old sledge.   He is a nice young man,
and puts himself out a good deal to make me comfortable.

"I found something yesterday at the table that bothered
me.   It was a new kind of a silver dingus, with two handles
to it, for getting a lump of sugar into your tea.   I saw right
away that it was for that, but when I took the two handles
in my hand like a nut cracker and tried to scoop up a lump
of sugar with it I felt embarrassed.   Several people who
were total strangers to me smiled.

"After dinner the waiter brought me a little pink-glass
bowl of lemonade and a clean wipe to dry my mouth with,

I reckon, after I drank the lemonade. I do not pine for lemonade much, anyhow, but this was specially poor. It was just plain water, with a lemon rind and no sugar into it.

"One rural rooster from Pittsburg showed his contempt for the blamed stuff by washing his hands in it. I may be rough and uncouth in my style, but I hope I will never lower myself like that in company."

O, The Man in the Moon has a crick in his back;
  Whee!
      Whimm!
         Ain't you sorry for him?
And a mole on his nose that is purple and black;
And his eyes are so weak that they water and run
If he dares to dream even he looks at the sun,—
So he just dreams of stars, as the doctors advise—
  My!
    Eyes!
      But isn't he wise—
To just dream of stars, as the doctors advise?
And The Man in the Moon has a boil on his ear—
  Whee!
    Whing!
      What a singular thing!
I know; but these facts are authentic, my dear,—
There's a boil on his ear, and a corn on his chin—
He calls it a dimple,—but dimples stick in—
Yet it might be a dimple turned over, you know;
  Whang!
    Ho!

Why, certainly so!—
It might be a͡ dimple turned over, you know!
And The Man in the Moon has a rheumatic knee—
Gee!
Whizz!
What a pity that is!
And his toes have worked round where his heels ought
to be.—
So whenever he wants to go North he goes South,
And comes back with porridge-crumbs all round his mouth,
And he brushes them off with a Japanese fan,
Whing!
Whann!
What a marvelous man!
What a very remarkably marvelous man!

his Christmas Sled.

I watch him, with his Christmas
    sled;
He hitches on behind
A passing sleigh, with glad
    hooray,
And whistles down the wind;
He hears the horses champ
    their bits,
And bells that jingle-jingle—
You Woolly Cap! you Scarlet Mitts!
You miniature " Kriss Kringle!"

I almost catch your secret joy—
    Your chucklings of delight,
The while you whizz where glory is
    Eternally in sight!
With you I catch my breath, as swift
    Your jaunty sled goes gliding
O'er glassy track and shallow drift,
    As I behind were riding!

He winks at twinklings of the frost,
    And on his airy race,
Its tingles beat to redder heat
    The rapture of his face:—
The colder, keener is the air,
    The less he cares a feather.
But, there! he's gone! and I gaze on
    The wintriest of weather!

Ah, boy! still speeding o'er the track
    Where none returns again,
To sigh for you, or cry for you,
    Or die for you were vain.—
And so, speed on! the while I pray
    All nipping frosts forsake you—
Ride still ahead of grief, but may
    All glad things overtake you!

Her Tired Hands.

BOARD a western train the other day, I held in my bosom for over seventy-five miles, the elbow of a large man whose name I do not know. He was not a railroad hog or I would have resented it. He was built wide and he couldn't help it, so I forgave him.

He had a large, gentle, kindly eye, and when he desired to spit, he went to the car door, opened it and decorated the entire outside of the train forgetting that our speed would help to give scope to his remarks.

Naturally as he sat there by my side, holding on tightly to his ticket and evidently afraid that the conductor would forget to come and get it, I began to figure out in my mind what might be his business. He had pounded one thumb so that the nail was black where the blood had settled under it. This might happen to a shoemaker, a carpenter, a blacksmith or most any one else. So it didn't help me out much, though it looked to me as though it might have been done by trying to drive a fence-nail through a leather hinge with the back of an axe and nobody but a farmer would try to do that. Following up the clue, I discovered that he had milked on his boots and then I knew I was right. The man who milks before daylight, in a dark barn, when the ther-

mometer is down to 28 degrees below and who hits his boot and misses the pail, by reason of the cold and the uncertain light and the prudishness of the cow, is a marked man. He cannot conceal the fact that he is a farmer unless he removes that badge. So I started out on that theory and remarked that this would pass for a pretty hard winter on stock.

The thought was not original with me, for I have heard it expressed by others either in this country or Europe. He said it would.

"My cattle has gone through a whole mowful o' hay sence October and eleven ton o' brand. Hay don't seem to have the goodness to it thet it hed last year, and with their new *pro*-cess griss mills they jerk all the juice out o'

brand, so's you might as well feed cows with excelsior and upholster your horses with hemlock bark as to buy brand."

"Well, why do you run so much to stock? Why don't you try diversified farming, and rotation of crops?"

"Well, probably you got that idee in the papers. A man that earns big wages writing Farm Hints for agricultural papers can make more money with a soft lead pencil and two or three season-cracked idees like that'n I can carrying of 'em out on the farm. We used to have a feller in the drugstore in our town that wrote such good pieces for the *Rural Vermonter* and made up such a good condition powder out of his own head, that two years ago we asked him to write a nessay for the annual meeting of the Buckwheat Trust, and to use his own judgment about choice of subject. And what do you s'pose he had selected for a nessay that took the whole forenoon to read?"

"What subject, you mean?"

"Yes."

"Give it up!"

"Well, he'd wrote out that whole blamed intellectual wad on the subject of 'The Inhumanity of Dehorning Hydraulic Rams.' How's that?"

"That's pretty fair."

"Well, farmin' is like runnin' a paper in regards to some things. Every feller in the world will take and turn in and tell you how to do it, even if he don't know a blame thing about it. There ain't a man in the United States to-day that don't secretly think he could run airy one if his other business busted on him, whether he knows the difference between a new milch cow and a horse hayrake or not. We had one of these embroidered night-shirt farmers come from town better'n three years ago. Been a toilet soap man and done well, and so he came out and bought a farm that had nothing to it but a fancy house and barn, a lot of medder in the front yard and a southern aspect. The farm was no good. You couldn't raise a disturbance on it. Well, what does he do? Goes and gits a passle of slim-tailed, yeller cows from New Jersey and aims to handle cream and diversified farming. Last year the cuss sent a load of cream over and tried to sell it at the new crematory while the funeral and hollercost was goin' on. I may be a sort of a chump myself, but I read my paper and don't get left like that."

"What are the prospects for farmers in your State?"

"Well, they are pore. Never was so pore, in fact, sence I've ben there. Folks wonder why boys leaves the farm. My boys left so as to get protected, they said, and so they went into a clothing-store, one of 'em, and one went into hardware and one is talking protection in the Legislature this winter. They said that farmin' was gittin' to be like fishin' and huntin', well enough for a man that has means and leisure, but they couldn't make a livin' at it, they said.

Another boy is in a drug store, and the man that hires him says he is a royal feller."

"Kind of a castor royal feller," I said, with a shriek of laughter.

He waited until I had laughed all I wanted to and then he said:

"I've always hollered for high terriff in order to hyst the public debt, but now that we've got the national debt coopered I wish they'd take a little hack at mine. I've put in fifty years farmin'. I never drank licker in any form. I've worked from ten to eighteen hours a day, been economical in cloze and never went to a show more'n a dozen times in my life, raised a family and learned upward of two hundred calves to drink out of a tin pail without blowing all their vittles up my sleeve. My wife worked alongside o' me sewin' new seats on the boys' pants, skimmin' milk and even helpin' me load hay. For forty years we toiled along together and hardly got time to look into each others' faces or dared to stop and get acquainted with each other. Then her health failed. Ketched cold in the spring house, prob'ly skimmin' milk and washin' pans and scaldin' pails and spankin' butter. Anyhow, she took in a long breath one day while the doctor and me was watchin' her, and she says to me, 'Henry,' says she, 'I've got a chance to rest,' and she put one tired, wore-out hand on top of the other tired, wore-out hand, and I knew she'd gone where they don't work all day and do chores all night.

"I took time to kiss her then. I'd been too busy for a good while previous to that, and then I called in the boys. After the funeral it was too much for them to stay around and eat the kind of cookin' we had to put up with, and nobody spoke up around the house as we used to. The boys quit whistlin' around the barn and talked kind of low by themselves about goin' to town and gettin' a job.

"They're all gone now and the snow is four feet deep on mother's grave up there in the old berryin' ground."

Then both of us looked out of the car window quite a long while without saying anything.

"I don't blame the boys for going into something else long's other things pays better; but I say — and I say what I know — that the man who holds the prosperity of this country in his hands, the man that actually makes money for other people to spend, the man that eats three good, simple, square meals a day and goes to bed at nine o'clock, so that future generations with good blood and cool brains can go from his farm to the Senate and Congress and the White House — he is the man that gets left at last to run his farm, with nobody to help him but a hired man and a high protective terriff. The farms in our State is mortgaged for over seven hundred million dollars. Ten of our Western States — I see by the papers — has got about three billion and a half mortgages on their farms, and that don't count the chattel mortgages filed with the town clerks on farm machinery, stock, waggins, and even crops, by gosh! that ain't two inches high under the snow. That's what the prospects is for farmers now. The Government is rich, but the men that made it, the men that fought perarie fires and perarie wolves and Injins and potato-bugs and blizzards, and has paid the war debt and pensions and everything else and hollered for the Union and the Republican party and free schools and high terriff and anything else that they was told to, is left high and dry this cold winter with a mortgage of seven billions and a half on the farms they have earned and saved a thousand times over."

"Yes; but look at the glory of sending from the farm the future President, the future Senator and the future member of Congress."

"That looks well on paper, but what does it really amount to? Soon as a farmer boy gits in a place like that he forgets the soil that produced him and holds his head as high as a holly-hock. He bellers for protection to everybody but the farmer, and while he sails round in a highty-tighty room with a fire in it night and day, his father on the farm has to kindle his own fire in the morning with elm slivvers, and he has to wear his son's lawn-tennis suit next to him or freeze to death, and he has to milk in an old gray shawl that has held that member of Congress when he was a baby, by gorry! and the old lady has to sojourn through the winter in the flannels that Silas wore at the riggatter before he went to Congress.

"So I say, and I think that Congress agrees with me, Damn a farmer, anyhow!"

He then went away.

Ezra House.

Come listen, good people, while a story I do tell,
Of the sad fate of one which I knew so passing well;
He enlisted at McCordsville, to battle in the south,
And protect his country's union; his name was Ezra House.

He was a young school-teacher, and educated high
In regards to Ray's arithmetic, and also Algebra.
He give good satisfaction, but at his country's call
He dropped his position, his Algebra and all.

"Its Oh, I'm going to leave you, kind scholars," he said—
For he wrote a composition the last day and read;
And it brought many tears in the eyes of the school,
To say nothing of his sweet-heart he was going to leave so
    soon.

" I have many recollections to take with me away,
Of the merry transpirations in the school-room so gay ;
And of all that's past and gone I will never regret
I went to serve my country at the first of the outset!"

He was a good penman, and the lines that he wrote
On that sad occasion was too fine for me to quote,—
For I was there and heard it, and I ever will recall
It brought the happy tears to the eyes of us all.

And when he left, his sweetheart she fainted away,
And said she could never forget the sad day
When her lover so noble, and gallant and gay,
Said " Fare you well, my true love!" and went marching
    away.

He hadn't been gone for more than two months
When the sad news come—" he was in a skirmish once,
And a cruel rebel ball had wounded him full sore
In the region of the chin, through the canteen he wore."

But his health recruited up, and his wounds they got well;
But while he was in battle at Bull Run or Malvern Hill,
The news come again, so sorrowful to hear—
" A sliver from a bombshell cut off his right ear."

But he stuck to the boys, and it's often he would write,
That " he wasn't afraid for his country to fight."
But oh, had he returned on a furlough, I believe
He would not, today, have such cause to grieve.

For in another battle—the name I never heard—
He was guarding the wagons when an accident occurred,—
A comrade, who was under the influence of drink,
Shot him with a musket through the right cheek, I think.

But his dear life was spared, but it hadn't been for long
'Till a cruel rebel colonel came riding along,
And struck him with his sword, as many do suppose,
For his cap-rim was cut off, and also his nose.

But Providence, who watches o'er the noble and the brave,
Snatched him once more from the jaws of the grave ;
And just a little while before the close of the war,
He sent his picture home to his girl away so far.

And she fell into decline, and she wrote in reply,
" She had seen his face again and was ready to die ;"
And she wanted him to promise, when she was in her tomb,
He would only visit that by the light of the moon.

But he never returned at the close of the war,
And the boys that got back said he hadn't the heart ;
But he got a position in a powder-mill, and said
He hoped to meet the doom that his country denied.

"Oh, Wilhelmina, Come Back!"

PERSONAL—Will the young woman who edited the gravy department and
corrected proof at our pie foundry for two days and then jumped the
game on the evening that we were to have our clergyman to dine with us,
please come back, or write to 32 Park Row, saying where she left the crackers
and cheese?

Come back, Wilhelmina, and be our little sunbeam once
more. Come back and cluster around our hearthstone at
so much per cluster.

If you think best we will quit having company at the
house, especially people who do not belong to your set.

We will also strive, oh so hard, to make it pleasanter
for you in every way. If we had known four or five years
ago that children were offensive to you, it would have been
different. But it is too late now. All we can do is to shut

them up in a barn and feed them through a knot-hole.   If
they shriek loud enough to give pain to your throbbing
brow, let no one know and we will overcome any false sentiment we may feel towards them and send them to the
Tombs.

Since you went away we can see how wicked and
selfish we were and how little we considered your comfort.
We miss your glad smile, also your Tennessee marble cake
and your slat pie.   We have learned a valuable lesson
since you went away, and it is that the blame should not
have rested on one alone.   It should have been divided
equally, leaving me to bear half of it and my wife the
other half.

Where we erred was in dividing up the blame on the
basis of tenderloin steak or peach cobbler, compelling you
to bear half of it yourself.   That will not work, Wilhelmina.
Blame and preserves do not divide on the same basis.   We
are now in favor of what may be called a sliding scale.   We
think you will like this better.

We also made a grave mistake in the matter of nights
out.   While young, I formed the wicked and pernicious
habit of having nights out myself.   I panted for the night
air and would go a long distance and stay out a long time
to get enough of it for a mess and then bring it home in a
paper bag, but I can see now that it is time for me to
remain indoors and give young people like yourself a
chance, Wilhelmina.

So, if I can do anything evenings while you are out that
will assist you, such as stoning raisins or neighboring windows command me.   I am no cook, of course, but I can
peel apples or grind coffee or hold your head for you when
you need sympathy.   I could also soon learn to do the
plain cooking, I think, and friends who come to see us after
this have agreed to bring their dinners.

There is no reason why harmony should not be restored
among us and the old sunlight come back to our roof tree.

Another thing I wish to write before I close this humiliating personal. I wish to take back my harsh and bitter words about your singing. I said that you sang like a shingle-mill, but I was mad when I said it, and I wronged you. I was maddened by hunger and you told me that mush and milk was the proper thing for a brain worker, and you refused to give me any dope on my dumpling. Goaded to madness by this I said that you sang like a shingle-mill, but it was not my better, higher nature that spoke. It was my grosser and more gastric nature that asserted itself, and I now desire to take it back. You do not sing like a shingle-mill; at least so much as to mislead a practiced ear.

Your voice has more volume, and when your upper register is closed, is mellower than any shingle-mill I ever heard.

Come back, Wilhemina. We need you every hour.

After you went away we tried to set the bread as we had seen you do it, but it was not a success. The next day it came off the nest with a litter of small, sallow rolls which would easily resist the action of acids.

If you cannot come back will you please write and tell me how you are getting along and how you contrive to insert air-holes into home-made bread?

A HINT OF SPRING.

'Twas but a hint of Spring
  —for still
The atmosphere was sharp
  and chill—
Save where the genial sunshine smote
The shoulders of my over-
  coat,
And o'er the snow beneath
  my feet
Laid spectral fences down
  the street.

My shadow even seemed
  to be
Elate with some new buoy-
  ancy,
And bowed and bobbed in
  my advance
With trippingest extravagance,
And when a bird sang out somewhere,
It seemed to wheel with me, and stare.

Above I heard a rasping stir —
And on the roof the carpenter
Was perched, and prodding rusty leaves
From out the choked and dripping eaves —
And some one, hammering about,
Was taking all the windows out.

Old scraps of shingles fell before
The noisy mansion's open door;
And wrangling children raked the yard,
And labored much, and laughed as hard,
And fired the burning trash I smelt
And sniffed again — so good I felt!

A Treat Ode

"Scurious-like," said the tree-toad,
"I've twittered fer rain all day;
    And I got up soon,
    And hollered till noon—
But the sun hit blazed away,
    Till I jest clumb down in a crawfish-hole,
    Weary at heart, and sick at soul!

"Dozed away fer an hour,
And I tackled the thing agin;
    And I sung, and sung,
    Till I knowed my lung
Was jest about give in;
    And then, thinks I, ef it don't rain now,
    There 're nothin' in singin' anyhow!

"Once in a while some farmer
Would come a driven' past;
    And he'd hear my cry,
    And stop and sigh—
Till I jest laid back, at last,
    And I hollered rain till I thought my throat
    Would bust wide open at ever' note!

" But I *fetched* her!—O I *fetched* her!—
'Cause a little while ago,
   As I kindo' set
   With one eye shet,
And a-singin' soft and low,
   A voice drapped down on my fevered brain
   Sayin',—' Ef you'll jest hush I'll rain! ' "

"Our Wife."

THE story opens in 1877, when, on an April morning, the yellow-haired "devil" arrived at the office of the Jack Creek *Pizenweed*, at 7 o'clock, and found the editor in. It was so unusual to find the editor in at that hour that the boy whistled in a low contralto voice, and passed on into the "news room," leaving the gentlemanly, genial and urbane editor of the *Pizenweed* as he had found him, sitting in his foundered chair, with his head immersed in a pile of exchanges on the table and his venerable Smith & Wesson near by, acting as a paper-weight. The gentlemanly, genial and urbane editor of the *Pizenweed* presented the appearance of a man engaged in sleeping off a long and aggravated case of drunk. His hat was on the back of his head, and his features were entirely obscured by the loose papers in which they nestled.

Later on, Elijah P. Beckwith, the foreman, came in, and found the following copy on the hook, marked "Leaded

Editorial," and divided it up into "takes" for the yellow-haired devil and himself:

"In another column of this issue will be found, among the legal notices, the first publication of a summons in an action for divorce, in which our wife is plaintiff and we are made defendant. While generally deprecating the practice of bringing private matters into public through the medium of the press, we feel justified in this instance, inasmuch as the summons sets forth, as a cause of action, that we are, and have been, for the space of ten years, a confirmed drunkard without hope of recovery, and totally unwilling to provide for and maintain our said wife.

"That we have been given to drink, we do not, at this time, undertake to deny or in any way controvert, but that we can not quit at any time, we do most earnestly contend.

"In 1867, on the 4th day of July, we married our wife. It was a joyful day, and earth had never looked to us so fair or so desirable as a summer resort as it did that day. The flowers bloomed, the air was fresh and exhilarating, the little birds and the hens poured forth their respective lays. It was a day long to be remembered, and it seemed as though we had never seen Nature get up and hump herself to be so attractive as she did on this special morning—the morning of all mornings—the morning on which we married our wife.

"Little did we then dream that after ten years of varying fortune we would to-day give utterance to this editorial, or that the steam power-press of the *Pizenweed* would squat this legal notice for divorce, *a vinculo et thoro*, into the virgin page of our paper. But such is the case. Our wife has abandoned us to our fate, and has seen fit to publish the notice in what we believe to be the spiciest paper published west of the Missouri river. It was not necessary that the notice should be published. We were ready at any time to admit service, provided that plaintiff would serve it while

we were sober. We can not agree to remain sober after ten
o'clock a. m. in order to give people a chance to serve notices
on us. But in this case plaintiff knew the value of adver-
tising, and she selected a paper that goes to the better
classes all over the Union. When our wife does anything
she does it right.

"For ten years our wife and we have trudged along together. It has been a record of errors and failures on our
part; a record of heroic devotion and forbearance on the
part of our wife. It is over now, and with nothing to remember that is not soaked full of bitterness and wrapped
up in red flannel remorse, we go forth to-day and herald
our shame by publishing to the world the fact, that as
husband, we are a depressing failure, while as a red-eyed
and a rum-soaked ruin and all-round drunkard, we are a
tropical triumph. We print this without egotism', and we
point to it absolutely without vain glory.

"Ah, why were we made the custodian of this fatal gift,
while others were denied? It was about the only talent we
had, but we have not wrapped it up in a napkin. Sometimes we have put a cold, wet towel on it, but we have
never hidden it under a bushel. We have put it out at
three per cent a month, and it has grown to be a thirst that
is worth coming all the way from Omaha to see. We do
not gloat over it. We do not say all this to the disparage-
ment of other bright, young drinkers, who came here at the
same time, and who had equal advantages with us. We do
not wish to speak lightly of those whose prospects for fill-
ing a drunkard's grave were at one time even brighter than
ours. We have simply sought to hold our position here in
the grandest galaxy of extemporaneous inebriates in the
wild and woolly West. We do not wish to vaunt our own
prowess, but we say, without fear of successful contradic-
tion, that we have done what we could.

"On the fourth page of this number will be found,

among other announcements, the advertisement of our wife
who is about to open up the old laundry at the corner of
Third and Cottonwood streets, in the Briggs building. We
hope that our citizens will accord her a generous patronage,
not so much on her husband's account, but because she is
a deserving woman, and a good laundress. We wish that
we could as safely recommend every advertiser who patron-
izes these columns as we can our wife.

"Unkind critics will make cold and unfeeling remarks
because our wife has decided to take in washing, and they
will look down on her, no doubt, but she will not mind it,
for it will be a pleasing relaxation to wash, after the ten
years of torch-light procession and Mardi Gras frolic she
has had with us. It is tiresome, of course, to chase a pil-
low case up and down the wash-board all day, but it is
easier and pleasanter than it is to run a one-horse Inebriate
Home for ten years on credit.

"Those who have read the *Pizenweed* for the past three
years will remember that it has not been regarded as an
outspoken temperance organ. We have never claimed that
for it. We have simply claimed that, so far as we are personally concerned, we could take liquor or we could let it
alone. That has always been our theory. We still make
that claim. Others have said the same thing, but were unable to do as they advertised. We have been taking it
right along, between meals for ten years. We now propose,
and so state in the prospectus, that we will let it alone.
We leave the public to judge whether or not we can do
what we claim."

After the foreman had set up the above editorial, he
went in to speak to the editor, but he was still slumbering.
He shook him mildly, but he did not wake. Then Elijah
took him by the collar and lifted him up so that he could
see the editor's face.

It was a pale, still face, firm in its new resolution to forever "let it alone." On the temple and under the heavy sweep of brown hair there was a powder-burned spot and the cruel affidavit of the "Smith and Wesson" that our wife had obtained her decree.

The editor of the *Pizenweed* had demonstrated that he could drink or he could let it alone.

My Bachelor Chum.

O a corpulent man is my
bachelor chum,
With a neck apoplectic and
thick,
And an abdomen on him as
big as a drum,
And a fist big enongh for
the stick;
With a walk that for grace is
clear out of the case,
And a wobble uncertain—
as though
His little bow-legs had for-
gotten the pace
That in youth used to favor
him so.

He is forty, at least; and the
top of his head
Is a bald and a glittering
thing;
And his nose and his two chubby cheeks are as red
As three rival roses in spring.
His mouth is a grin with the corners tucked in,
And his laugh is so breezy and bright
That it ripples his features and dimples his chin
With a billowy look of delight.

He is fond of declaring he "don't care a straw"—
    That "the ills of a bachelor's life
Are blisses compared with a mother-in-law,
    And a boarding-school miss for a wife!"
So he smokes, and he drinks, and he jokes and he winks,
    And he dines, and he wines all alone,
With a thumb ever ready to snap as he thinks
    Of the comforts he never has known.

But up in his den— (Ah my bachelor chum!)
    I have sat with him there in the gloom,
When the laugh of his lips died away to become
    But a phantom of mirth in the room!
And to look on him there you would love him, for all
    His ridiculous ways, and be dumb
As the little girl-face that smiles down from the wall
    On the tears of my bachelor chum.

The Philanthropical Jay.

It had been ten long weary years since I last met Jay Gould until I called upon him yesterday to renew the acquaintance and discuss the happy past. Ten years of patient toil and earnest endeavor on my part, ten years of philanthropy on his, have been filed away in the grim and greedy heretofore. Both of us have changed in that time, though Jay has changed more than I have. Perhaps that is because he has been thrown more in contact with change than I have.

Still, I had changed a good deal in those years, for when I called at Irvington yesterday Mr. Gould did not remember me. Neither did the watchful but overestimated dog in the front yard. Mr. Gould lives in comfort, in a cheery home, surrounded by hired help and a barbed-wire fence.

By wearing ready-made clothes, instead of having his clothing made especially for himself, he has been enabled to amass a good many millions of dollars with which he is enabled to buy things.

Carefully concealing the fact that I had any business relations with the press, I gave my card to the person who does chores for Mr. Gould, and, apologizing for not having dropped in before, I took a seat in the spare room to wait for the great railroad magnate.

Mr. Gould entered the room with a low, stealthy tread, and looked me over in a cursory way and yet with the air of a connoisseur.

"I believe that I have never had the pleasure of meeting you before, sir," said the great railroad swallower and amateur Philanthropist with a tinge of railroad irony.

"Yes, sir, we met some ten years ago," said I, lightly running my fingers over the keys of the piano in order to show him that I was accustomed to the sight of a piano. "I was then working in the rolling mill at Laramie City, Wyo., and you came to visit the mill, which was then operated by the Union Pacific Railroad Company. You do not remember me because I have purchased a different pair of trousers since I saw you, and the cane which I wear this season

changes my whole appearance also. I remember you, however, very much."

"Well, if we grant all that, Mr. Nye, will you excuse me for asking you to what I am indebted for this call?"

"Well, Mr. Gould," said I, rising to my full height and putting my soft hat on the brow of the Venus de Milo, after which I seated myself opposite him in a *degage* Western way, "you are indebted to *me* for this call. That's what you're indebted to. But we will let that pass. We are not here to talk about indebtedness, Jay. If you are busy you

needn't return this call till next winter. But I am here just
to converse in a quiet way, as between man and man; to
talk over the past, to ask you how your conduct is and to
inquire if I can do you any good in any way whatever. This

is no time to speak pieces and ask in a grammatical way,
'To what you are indebted for this call.' My main object
in coming up here was to take you by the hand and ask you
how your memory is this spring? Judging from what I
could hear, I was led to believe that it was a little inclined

to be sluggish and atrophied days and to keep you awake nights. Is that so, Jay?"

"No, sir; that is not so."

"Very well, then I have been misled by the reports in the papers, and I am glad it is all a mistake. Now, one thing more before I go. Did it ever occur to you that while you and your family are all out in your yacht together some day, a sudden squall, a quick lurch of the lee scuppers, a tremulous movement of the main brace, a shudder of the spring boom might occur and all be over?"

"Yes, sir. I have often thought of it, and of course such a thing might happen at any time; but you forget that while we are out on the broad and boundless ocean we enjoy ourselves. We are free. People with morbid curiosity cannot come and call on us. We cannot get the daily newspapers, and we do not have to meet low, vulgar people who pay their debts and perspire."

"Of course, that is one view to take of it; but that is only a selfish view. Supposing that you have made no provision for the future in case of accident, would it not be well for you to name some one outside of your own family to take up this great burden which is now weighing you down — this money which you say yourself has made a slave of you — and look out for it? Have you ever considered this matter seriously and settled upon a good man who would be willing to water your stock for you, and so conduct your affairs that nobody would get any benefit from your vast accumulations, and in every way carry out the policy which you have inaugurated?

"If you have not thoroughly considered this matter I wish that you would do so at an early date. I have in my mind's eye just such a man as you need. His shoulders are well fitted for a burden of this kind, and he would pick it up cheerfully at any time you see fit to lay it down. I will give you his address."

"Thank you," said Mr. Gould, as the thermometer in the next room suddenly froze up and burst with a loud report. "And now, if you will excuse me from offsetting my time, which is worth $500 a minute, against yours, which I judge to be worth about $1 per week, I will bid you good morning."

He then held the door open for me, and shortly after that I came away. There were three reasons why I did not remain, but the principal reason was that I did not think he wanted me to do so.

And so I came away and left him. There was little else that I could say after that.

It is not the first time that a Western man has been treated with consideration in his own section, only to be frowned upon and frozen when he meets the same man in New York.

Mr. Gould is below the medium height, and is likely to remain so through life. His countenance wears a crafty expression, and yet he allowed himself to be April-fooled by a genial little party of gentlemen from Boston, who salted the Central Branch of the Union Pacific Railroad by holding back all the freight for two weeks in order to have it on the road while Jay was examining the property.

Jay Gould would attract very little attention here on the streets, but he would certainly be looked upon with suspicion in Paradise. A man who would fail to remember that he had $7,000,000 that belonged to the Erie road, but who does not forget to remember whenever he paid his own hotel bills at Washington, is the kind of man who would pull up and pawn the pavements of Paradise within thirty days after he got there.

After looking over the above statement carefully, I feel called upon, in justice to myself, to state that Dr. Burchard did not assist me in constructing the last sentence.

For those boys who wish to emulate the example of Jay
Gould, the example of Jay Gould is a good example for
them to emulate.

If any little boy in New York on this beautiful Sabbath
morning desires to jeopardize his immortal soul in order to
be beyond the reach of want, and ride gayly over the
sunlit billows where the cruel fangs of the Excise law cannot reach him, let him cultivate a lop-sided memory, swap
friends for funds and wise counsel for crooked consols.

If I had thought of all this as I came down the front
steps at Irvington the other day, I would have said it to Mr.
Gould; but I did not think of it until I got home.   A man's
best thoughts frequently come to him too late for publication.

But the name of Jay Gould will not go down to future generations linked with those of Howard and Wilberforce. It will not go very far any way. In this age of millionaires, a millionaire more or less does not count very much, and only the good millionaires who baptize and beautify their wealth in the eternal sunlight of unselfishness will have any claim on immortality.

In this period of progress and high-grade civilization, when Satan takes humanity up to the top of a high mountain and shows his railroads and his kerosene oil and his distilleries and his coffers filled with pure leaf lard, and says: "All this will I give for a seat in the Senate," a common millionaire with no originality of design does not excite any more curiosity on Broadway than a young man who is led about by a little ecru dog.

I do not wish to crush capital with labor, or to further intensify the feeling which already exists between the two, for I am a land-holder and taxpayer myself, but I say that the man who never mixes up with the common people unless he is summoned to explain something and shake the moths out of his memory will some day, when the grass grows green over his own grave, find himself confronted by the same kind of a memory on the part of mankind.

I do not say all this because I was treated in an off-hand manner by Mr. Gould, but because I think it ought to be said.

As I said before, Jay Gould is considerably below the medium height and I am not going to take it back.

He is a man who will some day sit out on the corner of a new-laid planet with his little pink railroad maps on his knees, and ask "Where am I?" and the echoes from every musty corner of miasmatic oblivion will take up the question and refer it to the judiciary committee; but it will curl up and die like the minority report against a big railroad land grant.

"A Brave Refrain"

WHEN snow is here, and the trees look weird,
  And the knuckled twigs are gloved with frost;
When the breath congeals in the drover's beard,
  And the old pathway to the barn is lost;

When the rooster's crow is sad to hear,
  And the stamp of the stabled horse is vain,
And the tone of the cow-bell grieves the ear—
  O then is the time for a brave refrain!

When the gears hang stiff on the harness-peg,
  And the tallow gleams in frozen streaks;

And the old hen stands on a lonesome leg,
   And the pump sounds hoarse and the handle squeaks;
When the woodpile lies in a shrouded heap,
   And the frost is scratched from the window-pane,
And anxious eyes from the inside peep—
   O then is the time for a brave refrain!

When the ax-helve warms at the chimney-jamb!
   And hob-nailed boots on the hearth below,
And the house cat curls in a slumber calm,
   And the eight-day clock ticks loud and slow;
When the harsh broom-handle jabs the ceil
   'Neath the kitchen-loft, and the drowsy brain
Sniffs the breath of the morning meal—
   O then is the time for a brave refrain!

'Envoi.

When the skillet seethes, and a blubbering hot
Tilts the lid of the coffee-pot,
And the scent of the buckwheat cake grows plain—
O then is the time for a brave refrain!

A Blasted Snore.

Sleep, under favorable circumstances, is a great boon. Sleep, if natural and undisturbed, is surely as useful as any other scientific discovery. Sleep, whether administered at home or abroad, under the soporific influences of an underpaid preacher or the unyielding wooden cellar door that is used as a blanket in the sleeping car, is a harmless dissipation and a cheerful relaxation.

Let me study a man for the first hour after he has wakened and I will judge him more correctly than I would to watch him all winter in the Legislature. We think we are pretty well acquainted with our friends, but we are not thoroughly conversant with their peculiarities until we have seen them wake up in the morning.

I have often looked at the men I meet and thought what a shock it must be to the wives of some of them to wake up and see their husbands before they have had time to prepare, and while their minds are still chaotic.

The first glimpse of a large, fat man, whose brain has drooped down behind his ears, and whose wheezy breath wanders around through the catacombs of his head and then emerges from his nostrils with a shrill snort like the yelp of the damned, must be a charming picture for the eye of a delicate and beautiful second wife; one who loves to look on green meadows and glorious landscapes; one who has always wakened with a song and a ripple of laughter that fell on her father's heart like a shower of sunshine in the sombre green of the valley.

It is a pet theory of mine that to be pleasantly wakened is half the battle for the day. If we could be wakened by the refrain of a joyous song, instead of having our front teeth knocked out by one of those patent pillow-sham holders that sit up on their hind feet at the head of the bed, until we dream that we are just about to enter Paradise and have just passed our competitive examination, and which then swoop down and mash us across the bridge of the nose, there would be less insanity in our land and death would be regarded more in the light of a calamity.

When you waken a child do it in a pleasant way. Do not take him by the ear and pull him out of bed. It is disagreeable for the child, and injures the general *tout ensemble* of the ear. Where children go to sleep with tears on their cheeks and are wakened by the yowl of dyspeptic parents, they have a pretty good excuse for crime in after years. If I sat on the bench in such cases I would mitigate the sentence.

It is a genuine pleasure for me to wake up a good-natured child in a good-natured way. Surely it is better from those dimpled lids to chase the sleep with a caress than to knock out slumber with a harsh word and a bed slat.

No one should be suddenly wakened from a sound sleep. A sudden awaking reverses the magnetic currents, and makes the hair pull, to borrow an expression from Dante. The awaking should be natural, gradual, and deliberate.

A sad thing occurred last summer on an Omaha train. It was a very warm day, and in the smoking-car a fat man, with a magenta fringe of whiskers over his Adam's apple, and a light, ecru lambrequin of real camel's hair around the suburbs of his head, might have been discovered.

He could have opened his mouth wider, perhaps, but not without injuring the mainspring of his neck and turning his epiglottis out of doors.

He was asleep.

He was not only slumbering, but he was putting the earnestness and passionate devotion of his whole being into

it.   His shiny, oilcloth grip, with the roguish tip of a discarded collar just peeping out at the side, was up in the
iron wall-pocket of the car.   He also had, in the seat with
him, a market basket full of misfit lunch and a two-bushel
bag containing extra apparel.   On the floor he had a crock
of butter with a copy of the Punkville *Palladium* and *Stock
Grower's Guardian* over the top.

He slumbered on in a rambling sort of a way, snoring
all the time in monosyllables, except when he erroneously
swallowed his tonsils, and then he would struggle awhile
and get black in the face, while the passengers vainly hoped
that he had strangled.

While he was thus slumbering, with all the eloquence
and enthusiasm of a man in the full meridian of life, the
train stopped with a lurch, and the brakeman touched his
shoulder.

"Here's your town," he said.   "We only stop a minute.
You'll have to hustle."

The man, who had been far away, wrestling with Mor-
pheus, had removed his hat, coat, and boots, and when he
awoke his feet absolutely refused to go back into the same
quarters.

At first he looked around reproachfully at the people in
the car.   Then he reached up and got his oilcloth grip from
the bracket.   The bag was tied together with a string, and
as he took it down the string untied.   Then we all discovered that this man had been on the road for a long time,
with no object, apparently, except to evade laundries.   All
kinds of articles fell out in the aisle.   I remember seeing a
chest-protector and a linen coat, a slab of seal-brown ginger-
bread and a pair of stoga boots, a hairbrush and a bologna
sausage, a plug of tobacco and a porous plaster.

He gathered up what he could in both arms, made two
trips to the door and threw out all he could, tried again to put

his number eleven feet into his number nine boots, gave it up, and socked himself out of the car as it began to move, while the brakeman bombarded him through the window for two miles with personal property, groceries, dry-goods, boots

and shoes, gents' furnishing goods, hardware, notions, *bric-a-brac*, red herrings, clothing, doughnuts, vinegar bitters, and facetious remarks.

Then he picked up the retired snorer's railroad check from the seat, and I heard him say: "Why, dog on it, that wasn't his town after all."

Good-bye er Howdy-Do.

Say good-bye er howdy-do —
What's the odds betwixt the two?
Comin'— goin'— every day —
Best friends first to go away —
Grasp of hands you druther hold
Than their weight in solid gold,
Slips their grip while greetin' you.—
Say good-bye er howdy-do?

Howdy-do, and then, good-bye —
Mixes jest like laugh and cry;
Deaths and births, and worst and best
Tangled their contrariest;
Ev'ry jinglin' weddin'-bell
Skeearin' up some funeral knell.—
Here's my song, and there's your sigh:
Howdy-do, and then, good-bye!

Say good-bye er howdy-do —
Jest the same to me and you;
'Taint worth while to make no fuss,
'Cause the job's put up on us!
Some one's runnin' this concern
That's got nothin' else to learn —
If he's willin', we'll pull through.
Say good-bye or howdy-do!

SOCIETY GURGS from SANDY MUSH

The following constitute the items of great interest occurring on the East Side among the colored people of Blue Ruin:

Montmorency Tousley of Pizen Ivy avenue cut his foot badly last week while chopping wood for a party on Willow street. He has been warned time and again not to chop wood when the sign was not right, but he would not listen to his friends. He not only cut off enough of his foot to weigh three or four pounds, but completely gutted the coffee sack in which his foot was done up at the time. It will be some time before he can radiate around among the boys on Pizen avenue again.

Plum Beasley's house caught on fire last Tuesday night. He reckons it was caused by a defective flue, for the fire caught in the north wing. This is one of Plum's bon mots, however. He tries to make light of it, but the wood he has been using all winter was white birch, and when he got a big dose of hickory at the same place last week it was so dark that he didn't notice the difference and before he knew

it he had a bigger fire than he had allowed. In the midst
of a pleasant flow of conversation gas collected in the wood
and caused an explosion which threw a passel of live coals
on the bed. The house was soon a solid mass of flame.
Mr. Beasley is still short two children.

Mr. Granulation Hicks, of Boston, Mass., who has won
deserved distinction in advancing the interests of Sir
George Pullman, of Chicago, is here visiting his parents,
who reside on Upper Hominy. We are glad to see Mr.
Hicks and hope he may live long to visit Blue Ruin and
propitiate up and down our streets.

Miss Roseola Cardiman has just been the recipient of
a beautiful pair of chaste ear-bobs from her brother, who
is a night watchman in a jewelry store run by a man named
Tiffany in New York. Roseola claims that Tiffany makes
a right smart of her brother, and sets a heap by him.

Whooping-cough and horse distemper are again making
fearful havoc among the better classes at the foot of Pizen
Ivy avenue.

We are pained to learn that the free reading room,
established over Amalgamation Brown's store, has been
closed up by the police. Blue Ruin has clamored for a
free temperance reading room and brain retort for ten years,
and now a ruction between two of our best known citizens,
over the relative merits of a natural pair and a doctored
flush, has called down the vengeance of the authorities, and
shut up what was a credit to the place and a quiet resort,
where young men could come night after night and kind of
complicate themselves at. There are two or three men in
this place that will bully or bust everything they can get
into, and they have perforated more outrages on Blue Ruin
than we are entitled to put up with.

There was a successful doings at the creek last Sabbath,
during which baptism was administered to four grown

people and a dude from Sandy Mush.   The pastor thinks
it will take first-rate, though it is still too soon to tell.

Surrender Adams got a letter last Friday from his son
Gladstone, who filed on a homestead near Porcupine, Dak.,
two years ago.   He says they have had another of those
unprecedented winters there for which Dakota is so justly
celebrated.   He thinks this one has been even more so than
any of the others.   He wishes he was back here at Blue
Ruin, where a man can go out doors for half an hour without getting ostracized by the elements.   He says they brag
a good deal on their ozone there, but he allows that it can
be overdone.   He states that when the ozone in Dakota is
feeling pretty well and humping itself and curling up sheet-
iron roofs and blowing trains off the track, a man has to tie
a clothes-line to himself, with the other end fastened to the
door knob, before it is safe to visit his own hen-house.   He
says that his nearest neighbor is seventeen miles away, and
a man might as well buy his own chickens as to fool his
money away on seventeen miles of clothes-line.

It is a first-rate letter, and the old man wonders who
Gladstone got to write it for him.

The valuable ecru dog of our distinguished townsman,
Mr. Piedmont Babbit, was seriously impaired last Saturday
morning by an east-bound freight.

He will not wrinkle up his nose at another freight train.

George Wellington, of Hickory, was in town the front end
of the week.   He has accepted a position in the livery, feed
and sale stable at Sandy Mush.   Call again, George.

Gabriel Brant met with a sad mishap a few days since
while crossing the French Broad river, by which he lost his
leg.

Any one who may find an extra leg below where the
accident occurred will confer a favor on Mr. Brant by
returning same to No. 06½ Pneumonia street.   It may be

readily identified by any one, as it is made of an old pick-handle and weighs four pounds.

J. Quincy Burns has written a war article for the Century magazine, regarding a battle where he was at. In this article he aims to describe the sensations of a man who is ignorant of physical fear and yet yearns to have the matter submitted to arbitration. He gives a thorough expose of his efforts in trying to find a suitable board of arbitration as soon as he saw that the enemy felt hostile and eager for the fray.

The forthcoming number of the Century will be eagerly snapped up by Mr. Burns' friends who are familiar with his pleasing and graphic style of writing. He describes with wonderful power the sense of utter exhaustion which came over him and the feeling of bitter disappointment when he realized that he was too far away to participate in the battle and too fatigued to make a further search for suitable arbitrators.

"HE SMOKES—AND THAT'S ENOUGH," SAYS MA—

While Cigarettes to Ashes Turn.

I.

"He smokes—and that's enough," says Ma—
"And cigarettes, at that!" says Pa.

"He must not call again" says she—
"He *shall* not call again!" says he.

They both glare at me as before—
Then quit the room and bang the door,—

While I, their willful daughter, say,
"I guess I'll love him, anyway!"

II.

At twilight, in his room, alone,
His careless feet inertly thrown

Across a chair, my fancy can
But worship this most worthless man!

I dream what joy it is to set
His slow lips round a cigarette,

With idle-humored whiff and puff—
Ah! this is innocent enough!

To mark the slender fingers raise
The waxen match's dainty blaze,

Whose chastened light an instant glows
On drooping lids and arching nose,

Then, in the sudden gloom, instead,
A tiny ember, dim and red,

Blooms languidly to ripeness, then
Fades slowly, and grows ripe again.

III.

I lean back, in my own boudoir—
The door is fast, the sash ajar;

And in the dark, I smiling stare
At one wide window over there,

Where some one, smoking, pinks the gloom
The darling darkness of his room!

I push my shutters wider yet,
And lo! I light a cigarette;

And gleam for gleam, and glow for glow,
Each pulse of light a word we know,

We talk of love that still will burn
While cigarettes to ashes turn.

Says He.

"WHATEVER the weather may be,'
    says he—
  "Whatever the weather may be—
Its plaze, if ye will, an' I'll say me
    say—
Supposin' to-day was the winterest
    day,
Wud the weather be changing be-
    cause ye cried,
Or the snow be grass were ye cruci-
    fied?
The best is to make your own sum-
    mer," says he,
"Whatever the weather may be,"
    says he—
  "Whatever the weather may be!

"Whatever the weather may be,"
    says he—
  "Whatever the weather may be,
Its the songs ye sing, an' the smiles ye wear
That's a-makin' the sunshine everywhere;
An' the world of gloom is a world of glee,
Wid the bird in the bush, an' the bud in the tree,
Whatever the weather may be," says he—
"Whatever the weather may be!

" Whatever the weather may be," says he —
  " Whatever the weather may be,
  Ye can bring the spring, wid its green an˙ gold,
  An' the grass in the grove where the snow lies cold,
  An' ye'll warm your back, wid a smiling face,
  As ye sit at your heart like an owld fireplace,
  Whatever the weather may be," says he'
  " Whatever the weather may be! "

Where the Roads are Engaged in Forking.

I am writing this at an imitation hotel where the roads fork. I will call it the Fifth Avenue Hotel because the hotel at a railroad junction is generally called the Fifth Avenue, or the Gem City House, or the Palace Hotel. I stopped at an inn some years since called the Palace, and I can truly say that if it had ever been a palace it was very much run down when I visited it.

Just as the fond parent of a white-eyed, two-legged freak of nature loves to name his mentally-diluted son Napoleon, and for the same reason that a prominent horse owner in Illinois last year socked my name on a tall, buckskin-colored colt that did not resemble me, intellectually or physically, a colt that did not know enough to go around a barbed-wire fence, but sought to sift himself through it into an untimely grave, so this man has named his sway-backed wigwam the Fifth Avenue Hotel.

It is different from the Fifth avenue in many ways. In the first place there is not so much travel and business in its neighborhood. As I said before, this is where two rail-roads fork. In fact that is the leading industry here. The growth of the town is naturally slow, but it is a healthy growth. There is nothing in the nature of dangerous or wild-cat speculation in the advancement of this place, and while there has been no noticeable or rapid advance in the principal business, there has been no falling off at all and these roads are forking as much today as they did before the war, while the same three men who were present for the first glad moment are still here to witness the operation.

Sometimes a train is derailed, as the papers call it, and two or three people have to remain over as we did all night. It is at such a time that the Fifth Avenue Hotel is the scene of great excitement. A large cod-fish, with a broad and sunny smile and his bosom full of rock salt, is tied in the creek to freshen and fit himself for the responsible position of floor manager of the codfish ball.

A pale chambermaid, wearing a black jersey with large pores in it through which she is gently percolating, now goes joyously up the stairs to make the little post-office lock-box rooms look ten times worse than they ever did before. She warbles a low refrain as she nimbly knocks loose the venerable dust of centuries and sets it afloat throughout the rooms. All is bustle about the house.

Especially the chambermaid.

We were put in the guests' chamber here. It has two atrophied beds made up of pains and counterpanes.

This last remark conveys to the reader the presence of a light, joyous feeling which is wholly assumed on my part.

The door of our room is full of holes where locks have been wrenched off in order to let the coroner in. Last night I could imagine that I was in the act of meeting, personally, the famous people who have tried to sleep here and who moaned through the night and who died while waiting for the dawn.

I have no doubt in the world but there is quite a good-sized delegation from this hotel, of guests who hesitated about committing suicide, because they feared to tread the red-hot sidewalks of perdition, but who became desperate at last and resolved to take their chances, and they have never had any cause to regret it.

We washed our hands on doorknob soap, wiped them on a slippery elm court-plaster, that had made quite a reputation for itself under the nom-de-plume of "Towel," tried to warm ourselves at a pocket inkstand stove, that gave out heat like a dark lantern and had a deformed elbow at the back of it.

The chambermaid is very versatile, and waits on the table while not engaged in agitating the overworked mat-

tresses and puny pillows up-stairs. In this way she imparts the odor of fried pork to the pillow-cases and kerosene to the pie.

She has a wild, nervous and apprehensive look in her eye as though she feared that some herculean guest might seize her in his great strong arms and bear her away to a justice of the peace and marry her. She certainly cannot fully realize how thoroughly secure she is from such a calamity. She is just as safe as she was forty years ago, when she promised her aged mother that she would never elope with any one.

Still, she is sociable at times and converses freely with me at table, as she leans over my shoulder, pensively brushing the crumbs into my lap with a general utility towel, which accompanies her in her various rambles through the house, and she asks what we would rather have — "tea or eggs?"

This afternoon we will pay our bill, in accordance with a life-long custom of ours, and go away to permeate the busy haunts of men. It will be sad to tear ourselves away from the Fifth Avenue Hotel at this place; still, there is no great loss without some small gain, and at our next hotel we may not have to chop our own wood and bring it up stairs when we want to rest. The landlord of a hotel who goes away to a political meeting and leaves his guests to chop their own wood, and then charges them full price for the rent of a boisterous and tempest-tossed bed, will never endear himself to those with whom he is thrown in contact.

We leave at 2:30 this afternoon, hoping that the two railroads may continue to fork here just the same as though we had remained.

McFeeters' Fourth.

It was needless to say 'twas a glorious day,
And to boast of it all in that spread-eagle way
That our forefathers had since the hour of the birth
Of this most patriotic republic on earth!
But 'twas justice, of course, to admit that the sight
Of the old Stars-and-stripes was a thing of delight
In the eyes of a fellow, however he tried
To look on the day with a dignified pride
That meant not to brook any turbulent glee,
Or riotous flourish of loud jubilee!

So argued McFeeters, all grim and severe,
Who the long night before, with a feeling of fear,

Had slumbered but fitfully, hearing the swish
Of the sky-rocket over his roof, with a wish
That the urchin who fired it were fast to the end
Of the stick to forever and ever ascend;
Or to hopelesly ask why the boy with the horn
And its horrible havoc had ever been born!
Or to wish, in his wakefulness, staring aghast,
That this Fourth of July were as dead as the last!

So, yesterday morning, McFeeters arose,
With a fire in his eyes, and a cold in his nose,
And a gutteral voice in appropriate key
With a temper as gruff as a temper could be.
He growled at the servant he met on the stair,
Because he was whistling a national air,
And he growled at the maid on the balcony, who
Stood enrapt with the tune of "The Red White and Blue"
That a band was discoursing like mad in the street,
With drumsticks that banged, and with cymbals that beat.

And he growled at his wife, as she buttoned his vest,
And applausively pinned a rosette on his breast
Of the national colors, and lured from his purse
Some change for the boys—for firecrackers—or worse;
And she pointed with pride to a soldier in blue
In a frame on the wall, and the colors there, too;
And he felt, as he looked on the features, the glow
The painter found there twenty long years ago,
And a passionate thrill in his breast, as he felt
Instinctively round for the sword in his belt.

What was it that hung like a mist o'er the room?—
The tumult without—and the music—the boom
Of the cannon—the blare of the bugle and fife?—
No matter!—McFeeters was kissing his wife,

And laughing and crying and waving his hat
Like a genuine soldier, and crazy, at that!
—But it's needless to say 'twas a glorious day,
And to boast of it all in that spread-eagle way
That our forefathers have since the hour of birth
Of this most patriotic republic on earth!

In a Box.

I saw them last night in a box at the
    play—
  Old age and young youth side by side—
You might know by the glasses that
    pointed that way
  That they were—a groom and a bride;
And you might have known, too, by the
    face of the groom,

And the tilt of his head, and the grim
Little smile of his lip, he was proud to presume
That we men were all envying him.

Well, she was superb—an Elaine in the face,
A Godiva in figure and mien,
With the arm and the wrist of a Parian "Grace,"
And the high-lifted brow of a queen;
But I thought, in the splendor of wealth and of pride,
And in all her young beauty might prize,
I should hardly be glad if she sat by my side
With that far-away look in her eyes.

Seeking to Set the Public Right.

WOULD like to make an explanation at this time which concerns me, of course, more than any one else, and yet it ought to be made in the interests of general justice, also. I refer to a recent article published in a W e s t e r n paper and handsomely

illustrated in which, among others, I find the foregoing picture of my residence:

The description which accompanies the cut, among other things, goes on to state as follows: "The structure is elaborate, massive and beautiful. It consists of three stories, basement and attic, and covers a large area on the ground. It contains an elevator, electric bells, steam-heating arrangements, baths, hot and cold, in every room, electric lights, laundry, fire-escapes, &c. The grounds consist of at least five acres, overlooking the river for several miles up and down, with fine boating and a private fish-pond of two acres in extent, containing every known variety of game fish. The grounds are finely laid out in handsome

drives and walks, and when finished the establishment will
be one of the most complete and beautiful in the North-
west."

No one realizes more fully than I the great power of the
press for good or evil. Rightly used the newspaper can
make or unmake men, and wrongly used it can be even
more sinister. I might say, knowing this as I do, I want to
be placed right before the people. The above is not a
correct illustration or description of my house, for several
reasons. In the first place, it is larger and more robust in
appearance, and in the second place it has not the same
tout ensemble as my residence. My house is less obtrusive
and less arrogant in its demeanor than the foregoing and it
has no elevator in it.

My house is not the kind that seems to crave an elevator. An elevator in my house would lose money. There is
no popular clamor for one, and if I were to put one in I
would have to abolish the dining-room. It would also interfere with the parlor.

I have learned recently that the correspondent who
came here to write up this matter visited the town while
I was in the South, and as he could not find me he was at
the mercy of strangers. A young man who lives here and
who is just in the heyday of life, gleefully consented to show
the correspondent my new residence not yet completed.
So they went over and examined the new Oliver Wendell
Holmes Hospital, which will be completed in June and
which is, of course, a handsome structure, but quite different
from my house in many particulars.

For instance, my residence is of a different school of arch-
itecture, being rather on the Scandinavian order, while the
foregoing has a tendency toward the Ironic. The hospital
belongs to a very recent school, as I may say, while my
residence, in its architectural methods and conception, goes
back to the time of the mound builders, a time when a Gothic

hole in the ground was considered the *magnum bonum* and the scrumptuous thing in art.   If the reader will go around behind the above building and notice it carefully on the east side, he will not discover a dried coonskin nailed to the rear breadths of the wood-shed.   That alone ought to convince an observing man that the house is not mine.   The coonskin regardant will always be found emblazoned on my arms, together with a blue Goddess of Liberty and my name in green India ink.

Above I give a rough sketch of my house.   Of course I have idealized it somewhat, but only in order to catch the eye of the keenly observant reader. The front part of the house runs back to the time of Polypus the First, while the L, which does not show in the drawing, runs back as far as the cistern.

In closing, let me say that I am not finding fault with any one because the above error has crept into the public prints, for it is really a pardonable error, after all.   Neither do I wish to be considered as striving to eliminate my name from the columns of the press, for no one could be more tickled than I am over a friendly notice of my arrival in town or a timely reference to my courteous bearing and youthful appearance, but I want to see the Oliver Wendell Holmes Hospital succeed, and so I come out in this way over my own signature and admit that the building does not belong to me and that, so far as I am concerned, the man who files a lien on it will simply fritter away his time.

A Dose't of Blues.

I' got no patience with blues at all!
  And I ust to kindo talk
Aginst 'em, and claim, 'tel along last fall,
  They was none in the fambly stock;
But a nephew of mine, from Eelinoy,
  That visited us last year,
He kindo' convinct me different
  While he was a-stayin' here.

Frum ever'-which-way that blues is frum,
   They'd tackle him ever' ways;
They'd come to him in the night, and come
   On Sundys, and rainy days;
They'd tackle him in corn-plantin' time,
   And in harvest, an airly fall,
But a dose't of blues in the wintertime
   He 'lowed was the worst of all!

Said all diseases that ever he had—
   The mumps, er the rheumatiz—
Er ever-other-day aigger's bad
   Purt' nigh as anything is!—
Er a cyarbuncle, say, on the back of his neck,
   Er a fellon on his thumb,—
But you keep the blues away frum him,
   And all o' the rest could come!

And he'd moan, "They's nary a leaf below!
   Ner a spear o' grass in sight!
And the whole wood-pile's clean under snow!
   And the days is dark as night!
And you can't go out—ner you can't stay in—
   Lay down—stand up—ner set!"
And a case o' reguller tyfoid blues
   Would double him jest clean shet!

I writ his parents a postal-kyard
   He could stay 'tel spring-time come;
And Aprile first, as I rickollect,
   Was the day we shipped him home.
Most o' his relatives, sence then,
   Has either give up, er quit,
Er jest died off, but I understand
   He's the same old color yit!

Wanted, A Fox.

SLIPPERY ELMHURST,
STATEN ISLAND, July 18, 1888.

To the Editor:

DEAR SIR: Could you inform a constant reader of your valuable paper where he would be most likely to obtain a good, durable, wild fox which could be used for hunting purposes on my premises? I desire a fox that is a good roadster and yet not too bloodthirsty. If I could secure one that would not bite, it would tickle me most to death.

You know, perhaps, that I am of English origin. Some of the best and bluest blood of the oldest and most decrepit families in England flows in my veins. There is no better blood extant. We love the exhilarating sports of our ancesters, and nothing thrills us through and through like the free chase 'cross country behind the fleeing fox. Joyously we gallop over the sward behind the yelping pack, as we clearly scent high, low, jack and the game.

My ancestors are haughty English people from Piscataquis county, Maine. For centuries, our rich, warm, red blood has been mellowed by the elderberry wine and huckleberry juice of Moosehead lake; but now and then it will assert itself and mantle in the broad and indestructible cheek of our race. Ever and anon in our family you will notice the slender, triangular chest, the broad and haughty sweep of abdomen, and the high, intellectual expanse of pelvic bone, which denotes the true Englishman; proud, high-spirited, soaked full of calm disdain, wearing checked pantaloons, and a soft, flabby tourist's hat that has a bow at both ends, so that a man can not get too drunk to put it on his head wrong.

I know that here in democrat-ic America, where every man has to earn his living or marry rich, people will scorn my high-born love of the fox-chase, and speak in a slight-ing manner of my wild, wild yearn for the rush and scamper of the hunt. By Jove, but it is joy indeed to gallop over the sward and the cover, and the open land, the meet and the cucumber vines of the Plebeian farmer, to run over the wife of the peasant and tramp her low, coarse children into the rich mould, to "sick" the hounds upon the rude rustic as he paris greens his potatoes, to pry open the jaws of the pack and return to the open-eyed peasant the quiver-ing seat of his pantaloons, returning it to him not because it is lacking in its merit, but because it is not available.

Ah, how the pulses thrill as we bound over the lea, out across the wold, anon skimming the outskirts of the moor and going home with a stellated fracture of the dura mater through which the gas is gently escaping.

Let others rave over the dreamy waltz and the false joys

of the skating rink, but give me the maddening yelp of the pack in full cry as it chases the speckled two-year-old of the low born rustic across the open and into the pond.

Let others sing of the zephyrs that fan the white sails of their swift-flying yacht, but give me a wild gallop at the tail of my high-priced hounds and six weeks at the hospital with a fractured rib and I am proud and happy. All our family are that way. We do not care for industry for itself alone. We are too proud ever to become slaves to habits of industry. We can labor or we can let it alone.

This shows our superiority as a race. We have been that way for hundreds of years. We could work in order to be sociable, but we would not allow it to sap the foundations of our whole being.

I write, therefore, to learn, if possible, where I can get a good red or gray fox that will come home nights. I had a fox last season for hunting purposes, but he did not give satisfaction. He was constantly getting into the pound. I do not want an animal of that kind. I want one that I shall always know where I can put my hand upon him when I want to hunt.

Nothing can be more annoying than to be compelled to go to the pound and redeem a fox, when a party is mounted and waiting to hunt him.

I do not care so much for the gait of a fox, whether he lopes, trots or paces, so that his feet are sound and his wind good. I bought a light-red fox two years ago that had given perfect satisfaction the previous year, but when we got ready to hunt him he went lame in the off hind foot and crawled under a hen house back of my estate, where he remained till the hunt was over.

What I want is a young, flealess fox of the dark-red or iron-gray variety, that I can depend upon as a good road-ster ; one that will come and eat out of my hand and yearn to be loved.

I would like also a tall, red horse with a sawed-off tail; one that can jump a barbed wire fence without mussing it up with fragments of his rider.   Any one who may have such a horse or pipless fox will do well to communicate with me in person or by letter, enclosing references.   I may be found during the summer months on my estate, spread out under a tree, engaged in thought.

E. Fitzwilliam Nye.

Slipperyelmhurst, Staten Island, N. Y.

IMITATED.

Say! *you* feller! *You*—
   With that spade and
      the pick!—
What do you 'pose to
      do
   On this side o' the
      crick?
   Goin' to tackle this
   claim? Well I reckon
You'll let up agin purty quick!

No bluff, understand,—
   But the same has been tried,
And the claim never panned—
   Or the fellers has lied,—
For they tell of a dozen that tried it,
   And quit it most onsatisfied.

The luck's dead agin it!—
 The first man I see
That stuck a pick in it
 Proved *that* thing to me,—
For he˜sorto took down, and got homesick,
 And went back whar he'd orto' be!

Then others they worked it
 Some—more or less,
But finally shirked it,
 In grades of distress,—
With an eye out—a jaw or skull busted,
 Or some sort o' seriousness.

The *last* one was plucky—
 He wasn't afeerd,
And bragged he was "lucky,"
 And said that "he'd heerd
A heep of bluff-talk," and swore awkard
 He'd work any claim that he keered!

Don't you strike nary lick
 With that pick till I'm through;—
This-here feller talked slick
 And as peart-like as you!
And he says : "I'll abide here
 As long as I please!"

But he didn't..........He died here—
 And I'm his disease!

Seeking to be Identified.

CHICAGO, Feb. 20, 1888.

FINANCIAL circles here have been a good deal interested in the discovery of a cipher which was recently adopted by a depositor and which began to attract the attention at first of a gentleman employed in the Clearing-House. He was telling me about it and showing me the vouchers or duplicates of them.

It was several months ago that he first noticed on the back of a check passing through the Clearing-House the following cipher, written in a symmetrical, Gothic hand :

DEAR SIR : Herewith find payment for last month's butter. It was hardly up to the average. Why do you blonde your butter ? Your butter last month tried to assume an effeminate air, which certainly was not consistent with its great vigor. Is it not possible that this butter is the brother to what we had the month previous, and that it was exchanged for its sister by mistake ? We have generally liked your butter very much, but we will have to deal elsewhere if you are going to encourage it in wearing a full beard. Yours truly, W.

Moneyed men all over Chicago and financial cryptogrammers came to read the curious thing and to try and work out its bearing on trade. Everybody took a look at it and went away defeated. Even the men who were engaged in trying to figure out the identity of the Snell murderer, took a day off and tried their Waterbury thinkers on this problem. In the midst of it all another check passed through the Clearing-House with this cipher, in the same hand :

SIR : Your bill for the past month is too much. You forget the eggs returned at the end of second week, for which you were to give me credit. The cook broke one of them by mistake, and then threw up the portfolio of pie-founder in our once joyous home. I will not dock you for loss of cook, but I cannot allow you for the eggs. How you succeed in dodging quarantine with eggs like that is a mystery to yours truly, W.

Great excitement followed the discovery of this indorsement on a check for $32.87. Everybody who knew anything about ciphering was called in to consider it. A young man from a high school near here, who made a specialty of mathematics and pimples, and who could readily tell how long a shadow a nine-pound ground-hog would cast at 2 o'clock and 37 minutes P.M., on ground-hog day, if sunny, at the town of Fungus, Dak., provided latitude and longitude and an irregular mass of red chalk be given to him, was secured to jerk a few logarithms in the interests of trade. He came and tried it for a few days, covered the interior of the Exposition Building with figures and then went away.

The Pinkerton detectives laid aside their literary work on the great train book, entitled "The Jerkwater Bank Robbery and Other Choice Crimes," by the author of "How I Traced a Lame Man Through Michigan and Other Felonies." They grappled with the cipher, and several of them leaned up against something and thought for a long time, but they could make neither head nor tail to it. Ignatius Donnelly took a powerful dose of kumiss, and under its maddening influence sought to solve the great problem which threatened to engulf the national surplus. All was in vain. Cowed and defeated, the able conservators of coin, who require a man to be identified before he can draw on his overshoes at sight, had to acknowledge if this thing continued it threatened the destruction of the entire national fabric.

About this time I was calling at the First National Bank of Chicago, the greatest bank, if I am not mistaken, in America. I saw the bonds securing its issue of national currency the other day in Washington, and I am quite sure the custodian told me it was the greatest of any bank in the Union. Anyway, it was sufficient, so that I felt like doing my banking business there whenever it became handy to do so.

I asked for a certificate of deposit for $2,000, and had the money to pay for it, but I had to be identified. "Why," I said to the receiving teller, "surely you don't require a man to be identified when he deposits money, do you?"

"Yes, that's the idea."

"Well, isn't that a new twist on the crippled industries of this country?"

"No; that's our rule. Hurry up, please, and don't keep men waiting who have money and know how to do business."

"Well, I don't want to obstruct business, of course, but suppose, for instance, I get myself identified by a man I know and a man you know, and a man who can leave his business and come here for the delirious joy of identifying me, and you admit that I am the man I claim to be, corresponding as to description, age, sex, etc., with the man I advertise myself to be, how would it be about your ability to identify yourself as the man you claim to be? I go all over Chicago, visiting all the large pork-packing houses in search of a man I know, and who is intimate with literary people like me, and finally we will say I find one who knows me and who knows you, and whom you know, and who can leave his leaf lard long enough to come here and identify me all right. Can you identify yourself in such a way that when I put in my $2,000 you will not loan it upon insufficient security as they did in Cincinnati the other day, as soon as I go out of town?"

"Oh, we don't care especially whether you trade here or not, so that you hurry up and let other people have a chance. Where you make a mistake is in trying to rehearse a piece here instead of going out to Lincoln Park or somewhere in a quiet part of the city. Our rules are that a man who makes a deposit here must be identified.

"All right. Do you know Queen Victoria?"

"No, sir; I do not."

"Well, then, there is no use in disturbing her. Do you know any of the other crowned heads?"

"No, Sir."

"Well, then, do you know President Cleveland, or any of the Cabinet, or the Senate or members of the House?"

"No."

"That's it, you see. I move in one set and you in another. What respectable people do you know?"

"I'll have to ask you to stand aside, I guess, and give that string of people a chance. You have no right to take up my time in this way. The rules of the bank are inflexible. We must know who you are, even before we accept your deposit."

I then drew from my pocket a copy of the Sunday WORLD, which contained a voluptuous picture of myself. Removing my hat and making a court salaam by letting out four additional joints in my lithe and versatile limbs, I asked if any further identification would be necessary.

Hastily closing the door to the vault and jerking the

combination, he said that would be satisfactory.   I was
then permitted to deposit in the bank.

I do not know why I should always be regarded with
suspicion wherever I go.   I do not present the appearance

of a man who is steeped in crime, and yet when I put my
trivial little two-gallon valise on the seat of a depot waiting-
room a big man with a red mustache comes to me and hisses
through his clinched teeth: "Take yer baggage off the
seat!!"   It is so everywhere.   I apologize for disturbing a
ticket agent long enough to sell me a ticket, and he tries to
jump through a little brass wicket and throttle me.   Other

men come in and say: "Give me a ticket for Bandoline, O., and be dam sudden about it, too," and they get their ticket and go aboard the car and get the best seat, while I am begging for the opportunity to buy a seat at full rates and then ride in the wood box.   I believe that common courtesy and

decency in America need protection.   Go into an hotel or a hotel, whichever suits the eyether and nyether readers of these lines, and the commercial man who travels for a big sausage-casing house in New York has the bridal chamber, while the meek and lowly minister of the Gospel gets a wall-pocket room with a cot, a slippery-elm towel, a cake of cast-iron soap, a disconnected bell, a view of the laundry, a tin roof and $4 a day.

But I digress.   I was speaking of the bank check cipher.

At the First National Bank I was shown another of these remarkable indorsements. It read as follows:

DEAR SIR. This will be your pay for chickens and other fowls received up to the first of the present month. Time is working wondrous changes in your chickens. They are not such chickens as we used to get of you before the war. They may be the same chickens, but oh! how changed by the lapse of time! How much more indestructible! How they have learned since then to defy the encroaching tooth of remorseless ages, or any other man! Why do you not have them tender like your squashes? I found a blue poker chip in your butter this week. What shall I credit myself for it? If you would try to work your butter more and your customers less it would be highly appreciated, especially by, yours truly   W.

Looking at the signature on the check itself, I found it to be that of Mrs. James Wexford, of this city. Knowing Mr. Wexford, a wealthy and influential publisher here, I asked him to-day if he knew anything about this matter. He said that all he knew about it was that his wife had a separate bank account, and had asked him several months ago what was the use of all the blank space on the back of a check, and why it couldn't be used for correspondence with the remittee. Mr. Wexford said he'd bet $500 that his wife had been using her checks that way, for he said he never knew of a woman who could possibly pay postage on a note, remittance or anything else unless every particle of the surface had been written over in a wild, delirious, three-story hand. Later on I found that he was right about it. His wife had been sassing the grocer and the butter-man on the back of her checks. Thus ended the great bank mystery.

I will close this letter with a little incident, the story of which may not be so startling, but it is true. It is a story of child faith. Johnny Quinlan, of Evanston, has the most wonderful confidence in the efficacy of prayer, but he thinks that prayer does not succeed unless it is accompanied with considerable physical strength. He believes that adult prayer is a good thing, but doubts the efficacy of juvenile prayer.

He has wanted a Jersey cow for a good while and tried prayer, but it didn't seem to get to the central office. Last week he went to a neighbor who is a Christian and believer in the efficacy of prayer, also the owner of a Jersey cow.

"Do you believe that prayer will bring me a yaller Jersey cow?" said Johnny.

"Why, yes, of course. Prayer will remove mountains. It will do anything."

"Well, then, suppose you give me the cow you've got and pray for another one."